SPRINGER
LAB MANUAL

AF294885

Springer-Verlag Berlin Heidelberg GmbH

R. Prasad (Ed.)

Manual on Membrane Lipids

With 46 Figures

 Springer

Prof. Dr. Rajendra Prasad
Jawaharlal Nehru University
School of Life Sciences
110067 New Delhi
India

Library of Congress Cataloging-in-Publication
Manual on membrane lipids [edited by R. Prasad].
p. cm. – (Springer lab manual)
Includes bibliographical references
ISBN 978-3-642-48970-9   ISBN 978-3-642-79837-5 (eBook)
DOI 10.1007/978-3-642-79837-5

1. Membrane lipids – Laboratory manuals. I. Prasad, R.
(Rajendra), 1947- . II. Series.
QP752.M45M36 1996
95-25142
574.87'5–dc20

This work is subject to copyright. All rights are reserved, whether the whole or part of the material is concerned, specifically the rights of translation, reprinting, reuse of illustrations, recitation, broadcasting, reproduction on microfilm or in any other way, and storage in data banks. Duplication of this publication or parts thereof is permitted only under the provisions of the German Copyright Law of September 9, 1965, in its current version, and permissions for use must always be obtained from Springer-Verlag. Violations are liable for prosecution under the German Copyright Law.

Product liability: The publishers cannot guarantee the accuray of any information about dosage and application contained in this book. In every individual case the user must check such information by consulting the relevant literature.

© Springer-Verlag Berlin Heidelberg 1996
Originally published by Springer-Verlag Berlin Heidelberg New York in 1996

The use of general descriptive names, registered names, trademarks, etc. in this publication does not imply, even in the absence of a specific statement, that such names are exempt from the relevant protective laws and regulations and therefore free for general use.
Cover design: Struve & Partner, Heidelberg
Typesetting:Best-Set Typesetters, Hong Kong
SPIN 10123274 39/3137-5 4 3 2 1 0 - Printed on acid-free paper

Preface

Membrane research has reached a point where interdisciplinary collaboration is the only pragmatic strategy to resolve the mysteries of structure and function of supermolecular membrane organisation. The efforts of membrane biologists in the last few decades have concentrated in characterising the two major components of the membrane, i.e. protein and lipids, encompassing fields like pure chemistry, biochemistry and biophysics. It should, however, be acknowledged that membrane proteins have received considerably more attention compared with membrane lipids. This was partly owing to the fact that initially, the role of membrane lipids was thought only to be the provision of the hydrophobic core of the lipid bilayer. The reason for the presence of a variety of membrane lipids in a lipid bilayer was frequently questioned, especially when one type of lipid, e.g. phosphatidylcholine, alone can form a lipid bilayer. Such questions have since been answered. The discovery of a signal transduction mechanism, involving not only polyphosphatidylinositides but also phosphatidylcholines and sphingomyelins etc., has opened an entirely new dimension. In addition, several other emerging lines of evidence concerning roles of membrane lipids are providing further evidence for their multifaceted characteristics. These recognitions led to a renewed interest in lipid research, and have resulted in the publication of several excellent volumes dealing with the various and diverse roles of lipids. However, none of these is methodologically oriented to study membrane lipids. The present volume is a systematic attempt to fill that gap, and has therefore been carefully planned and presented as a "user-friendly" book. The protocol of a given method is preceded by the relevant background of each method, and is described in such a manner that it can be adapted to most laboratory situations.

The final selection of the methods routinely used to study membrane lipids was extremely difficult, since no manual can claim to contain all methods employed to study membrane lipid behaviour. This manual provides only a selection of the methods which are frequently followed in laboratories to study structure and function of membrane lipids. All the methods presented in this manual have been contributed by investigators who use them routinely. The protocols described here also include minor details to help researchers. Wherever possible, various tips, tricks and precautions etc. are discussed. In addition, certain protocols are also highlighted at relevant places in order to emphasise the instructions or tips at that particular step.

I have included a brief introductory chapter (Chap. I) on lipids, so that the researcher keen to follow the suggested methods has easy access to the details of common membrane lipids. Before identifying and characterising membrane lipids, the first prerequisite is to have pure membrane fractions as starting material. Chapter II deals exclusively with some well-tried methods of membrane isolation. The limited availability of the pure membrane fraction has been the biggest handicap, and as a result, in several instances lipids of pure membrane fractions are still not known. Obviously, isolation methods of all subcellular membranes of different origin could not be listed. Chapter II describes only selected protocols of isolation of a few types of membranes. One must, however, consult relevant references, especially in Methods in Enzymology, for different methods of isolation of membranes. Chapters III and IV describe step by step instructions to isolate, identify and quantify the membrane lipids. Again, protocols listed in these chapters are described only as examples which can be adapted to individual needs. Chapters V to IX deal with common approaches used to study asymmetry, phase transition, liposome formation and reconstitution, turnover and signal transduction of membrane lipids. These chapters are written by an expert in the respective field. As far as possible, methods have been described in steps that can be easily followed. However, in cases where this has not been possible, the method is discussed in the form of advice and examples. It is sincerely hoped that this volume will help those who are seeking common methods of membrane lipid isolation, characterisation, and the study of membrane phenomena, compiled in a single manual.

Spring 1995

Rajendra Prasad
New Delhi

Acknowledgements

My deepest gratitude is to my wife, Vibha, whose endurance has no parallel. She has been very supportive of this project, and very generous in allowing me the time it required. I am grateful to all my students, especially Pranab and Anjni, for offering me their unqualified support during the preparation of various drafts of this Volume. Without them, it would have been impossible to meet this academic and physical challenge. I am indebted to Ali Bhai, Jyotsna, Archana, Sanjeev, Krishnamurthy, Deepa and Raj Kishore, who read part of the text and offered helpful suggestions. I would also like to thank all my contributors who readily helped in producing the manual, and translating the concept into reality. Finally, I acknowledge the countless fond memories of my darling daughter, Juhi, who suddenly left us exactly two years ago, but who remains a constant companion and encourages me to keep going.

Rajendra Prasad   Spring 1995
New Delhi

Contents

Chapter VI Lipid Asymmetry of Membranes

List of Contributors

J. BOLARD
Laboratoire de Physique et
Chimie Biomoleculaires
CNRS UA 198,
Université Pierre et Marie
Curie,
4, Place Jussieu, 75252, Paris,
France

M. GHANNOUM
Harbor-UCLA Medical
Center,
1000 Carson Street,
Torrance, CA 90509,
USA

C.M. GUPTA
Institute of Microbial Tech-
nology,
Chandigarh-160014,
India

BASIL O. IBE
Department of Pediatrics
Harbor-UCLA,
RB-1, 1124 West Carson St.,
Torrance, CA 90509,
USA

ASHRAF S. IBRAHIM
Harbor-UCLA Medical
Center
1000 Carson Street,
Torrance, CA 90509,
USA

V.K. KALRA
Department of Biochemistry
and Molecular Biology,
USC School of Medicine,
2011 Zonal Avenue,
Los Angeles, CA 90033,
USA

A. KOUL
School of Life Sciences,
Jawaharlal Nehru University,
New Delhi-110067,
India

T.V. KULAKOVSKAYA
Institute of Biochemistry and
Physiology of
Microorganism, Russian
Academy of Sciences,
Pushchino,
Russian Federation

L.P. LICHKO
Institute of Biochemistry and
Physiology of
Microorganism, Russian
Academy of Sciences,
Pushchino,
Russian Federation

BÉLA NÉMET
Department of Physics,
Janus Pannonius University,
H-7624 Pécs, Ifjúság U 6,
Hungary

M.S. ODINTSOVA
Bach Institute of Biochemistry, Russian Academy
of Sciences, Moscow-117071,
Russian Federation

LEV A. OKOROKOV
Institute of Biochemistry and
Physiology of
Microorganism, Russian
Academy of Sciences,
Pushchino,
Russian Federation

M. OLLIVON
Laboratoire de Physique et
Chimie Biomoleculaires
CNRS UA 198,
Université Pierre et Marie
Curie,
4, Place Jussieu, Paris,
France

TIBOR PÁLI
Institute of Biophysics,
Biol. Res. Cent.,
Hungarian Academy of
Sciences,
PO Box 521,
H-6701 Szeged, Temesvakrit,
62,
Hungary

M. PATERNOSTRE
Equipe Physicochimie des
Systemes Polyphases,
URA CNRS 1218,
Universite Paris Sud.,
5 rue Jean-Baptiste Clement,
92296 Chatenay Malabry,
France

M. PÉSTI
Department of Botany,
Group of Microbiology,
Janus Pannonius University,
H-7624 Pecs, Ifjusag U 6,
Hungary

T. POMORSKI
Humboldt Universität zu
Berlin,
Mathematisch-
Naturwissenschaftliche
Fakultät 1,
Institut für Biologie/
Biophysik;
Invalidenstrasse 43,
10115 Berlin,
Germany

R. PRASAD
School of Life Sciences,
Jawaharlal Nehru University,
New Delhi-110067,
India

P.S. SASTRY
Department of Biochemistry,
Indian Institute of Science,
Bangalore-560012,
India

A. ZACHOWSKI
Institut de Biologie Physico-
Chimique,
13, rue Pierre et Marie Curie,
75005 Paris,
France

R.A. ZVYAGILSKAYA
Bach Institute of Biochemis-
try,
Russian Academy of Sci-
ences,
Moscow-117071,
Russian Federation

V.P. YURINA
Bach Institute of Bio-
chemistry,
Russian Academy of Sci-
ences,
Moscow-117071,
Russian Federation

Rajendra Prasad

1   Background

This chapter briefly presents some basic information on the structure and properties of membrane lipids. Since such information is already available in numerous biochemistry text books (Jain 1988; Zubay 1988; Lehninger et al. 1993), detailed discussion is avoided. This chapter provides ready reference to various common membrane lipids whose separation and identification is described in this manual. Typically, most of the lipids of an organism are found predominantly in membranes. Only a small amount of it is found elsewhere in cells. Table 1 depicts lipid composition of different membrane preparations. The data of Table 1 give some idea about variations and peculiarities of lipid composition of different membranes.

For discussion, lipids can be classified into two major classes: saponifiable (which contain fatty acids which form soaps, i.e. salts of fatty acids on alkaline hydrolysis) and non-saponifiable lipids (which do not contain fatty acids). The saponifiable lipids differ in their backbone structure, to which fatty acids are covalently attached. Structure and nomenclature of fatty acids are introduced before the different classes of lipids which contain them, are described.

School of Life Sciences, Jawaharlal Nehru University, New Delhi-110067, India

Table 1. Lipid composition of membrane preparations. (After Jain 1988)

Source	Percentage lipid									
	Cholesterol	PC[a]	SM[b]	PE[c]	PI[d]	PS[e]	PG[f]	DPG[g]	PA[h]	Glycolipids
Rat liver										
Plasma membrane	20	64		17	11		2			
Endoplasmic reticulum (rough)	6	55	3	16	8	3	–	–	–	
Endoplasmic reticulum (smooth)	10	55	12	21	6.7			1.9		
Mitochondria (inner)	<3	45	2.5	25	6	1	2	18	0.7	
Mitochondria (outer)	<5	50	5	23	13	2	2.5	3.5	1.3	
Nuclear	10	55	3	20	7	3	–	–	1	
Golgi	7.5	40	10	15	6	3.5	–	–	–	
Lysosomes	14	25	24	13	7	7	–	5	–	
Rat brain										
Myelin	22	11	6	14		7				
Synaptosome	20	24	3.5	20	2	8			1	21
Erythrocyte	24	31	8.5	15	2.2	7			<0.1	
Mycoplasma	0	–								
E. coli	0	–		80			15	5		
B. subtilis	0	0		30			12			
Chloroplast	0	4			1.5		6			55
Sindbis virus	0	26	18	35		20				

[a] Phosphatidylcholine.
[b] Sphingomyelin.
[c] Phosphatidylethanolamine.
[d] Phosphatidylinositol.
[e] Phosphatidylserine.
[f] Phosphatidylglycerol.
[g] Diphosphatidylglycerol.
[h] Phosphatidic acid.

2   Fatty Acids

All saponifiable lipids contain fatty acids as building blocks; however, only traces of fatty acids are present in free form in tissues and cells. Fatty acids are carboxylic acids with hydrocarbon chain and a terminal carboxyl group. The hydrocarbon chain may contain one (monoenoic or monounsaturated) or more than one (polyenoic or polyunsaturated) double bond. In certain instances, fatty acids could be cyclic and branched. Table 2 lists some of the commonly occurring fatty acids. While over 100 types of fatty acids exist in nature, palmitic acid ($C_{16:0}$), stearic acid ($C_{18:0}$) among saturated and oleic acid ($C_{18:1}$) among unsaturated are most common. In most of the fatty acids of higher organism, the first double bond is between carbon atoms 9 and 10. In polyunsaturated fatty acids, one double bond is between carbon atoms 9 and 10, while other double bonds usually occur between the 9, 10 double bond and the methyl terminal end of the chain. The configuration of double bond is mostly *cis-*, which provides a rigid kink of 30° in the hydrocarbon chain, whereas the *trans-* configuration results in an extended chain structure (Fig. 1).

The fatty acids differ from each other with respect to their hydrocarbon chain length (C_4–C_{36}), and degree of unsaturation (number of double bonds) which also govern their physical properties and of the compound that contains them. The melting point of fatty acids is strongly influenced by the length and degree of unsaturation (Table 2).

The following section very briefly describes some of the common saponifiable membrane lipids. Among nonsaponifiable lipids only sterols are mentioned.

Figure 2 shows various building blocks of some of the membrane lipids (saponifiable) which are briefly discussed in the following section.

Table 2. Some common fatty acids

Trivial name	Systematic name	mp[a]	Carbon skeleton	Structure
Saturated, straight chain fatty acids				
General formula: $CH_3(CH_2)_n COOH$				
Capric acid	n-Decanoic acid		10:0	$CH_3(CH_2)_8 COOH$
Lauric acid	n-Dodecanoic acid	44.2	12:0	$CH_3(CH_2)_{10} COOH$
Myristic acid	n-Tetradecanoic acid	53.9	14:0	$CH_3(CH_2)_{12} COOH$
Palmitic acid	n-Hexadecanoic acid	63.1	16:0	$CH_3(CH_2)_{14} COOH$
Stearic acid	n-Octadecanoic acid	69.6	18:0	$CH_3(CH_2)_{16} COOH$
Arachidic acid	n-Eicosanoic acid	76.5	20:0	$CH_3(CH_2)_{18} COOH$
Behenic acid	n-Docosanoic acid		22:0	$CH_3(CH_2)_{20} COOH$
Lignoceric acid	n-Tetracosanoic acid	86.0	24:0	$CH_3(CH_2)_{22} COOH$
Cerotic acid	n-Hexacosanoic acid		26:0	$CH_3(CH_2)_{24} COOH$
Unsaturated fatty acids				
Palmitoleic acid	cis-9-Hexadecenoic acid	−0.5	$16:1^{\Delta 9}$	$CH_3(CH_2)_5 CH = CH(CH_2)_7\text{-}COOH$
Oleic acid	cis-9-Octadecenoic acid	13.4	$18:1^{\Delta 9}$	$CH_3(CH_2)_7 CH = CH(CH_2)_7\text{-}COOH$
Vaccenic acid	cis-11-Octadecenoic acid		$18:1^{\Delta 11}$	$CH_3(CH_2)_5 CH = CH(CH_2)_9\text{-}COOH$
Linoleic acid	cis,cis-9,12-Octadecadienoic acid	−5.0	$18:2^{\Delta 9,12}$	$CH_3(CH_2)_4 CH = CHCH_2 CH = CH\text{-}$ $\text{-}CH(CH_2)_7 COOH$
α-Linolenic acid	All-cis-9,12,15-octadecatrienoic acid	−11	$18:3^{\Delta 9,12,15}$	$CH_3 CH_2 CH = CHCH_2 CH = CHCH_2\text{-}$ $\text{-}CH = CH(CH_2)_7 COOH$
Arachidonic acid	All-cis-5,8,11,14-eicosatetraenoic acid	−49.5	$20:4^{\Delta 5,8,11,14}$	$CH_3(CH_2)_4 CH = CHCH_2 CH = CH\text{-}$ $\text{-}CH_2 CH = CHCH_2 CH = CH(CH_2)_3 COOH$

[a] melting point (°C)

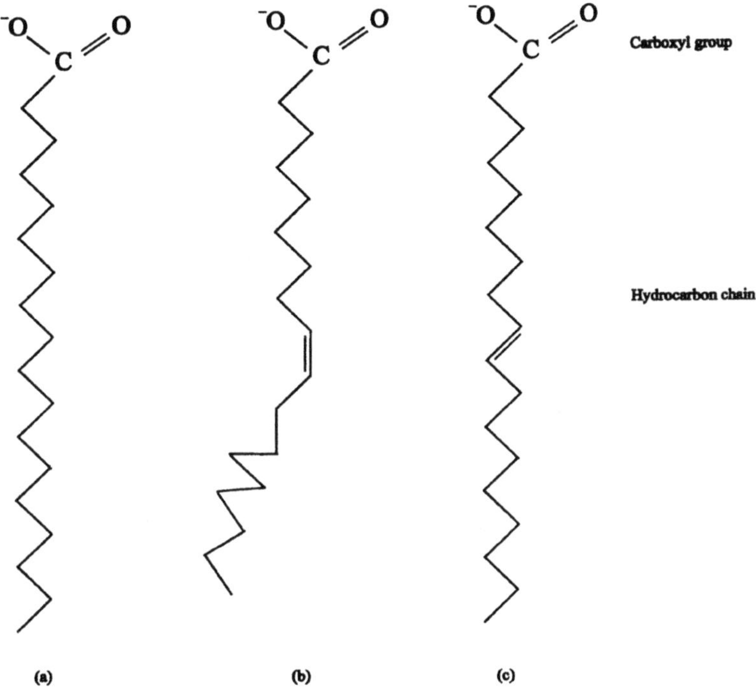

Fig. 1. Schematic representation of a saturated, 18:0; b, c unsaturated, 18:1$^{\Delta 9}$ fatty acids. The *cis* configuration of the double bond in b illustrates the characteristic rigid kink. The *trans* configuration of the double bond gives the most stable fully extended form c which is similar to that found in saturated fatty acids

3 Phosphoglycerides

These are the major component of membrane lipids. This class is loosely referred to as phospholipids/phosphatides, but not all phospholipids/ phosphatides contain glycerol, e.g. sphingomyelin is a phospholipid because it contains phosphorus (discussed later); however, it is classified as a sphingolipid owing to its backbone structure (sphingosine) to which fatty acid is linked. In phosphoglycerides, the terminal hydroxyl group of glycerol is esterified to phosphoric acid and the other

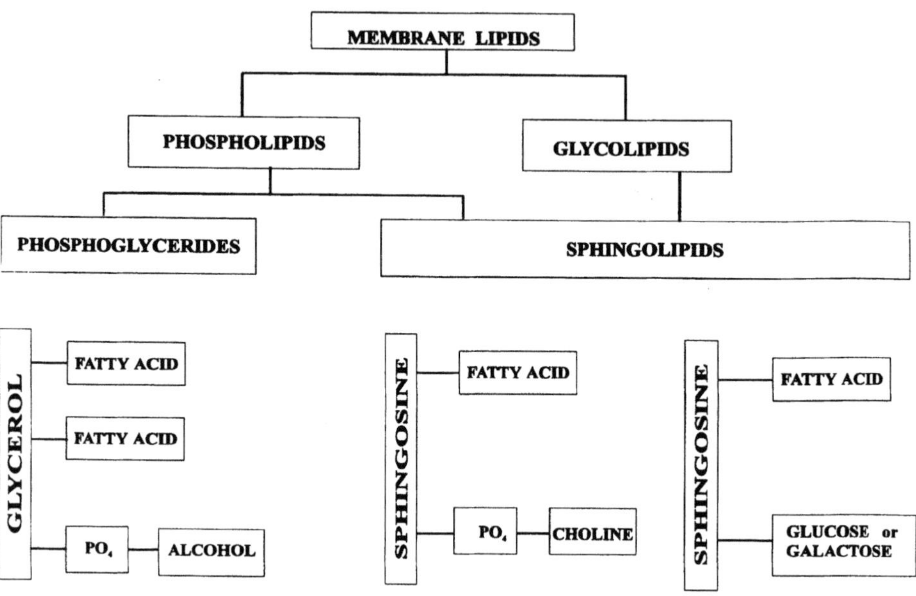

Fig. 2. Building blocks of phosphoglycerides and glycolipids

two are esterified to fatty acids. The resulting compound phosphatidic acid (PA) is common to all phosphoglycerides (Fig. 3). PA does not contain any polar head group (X). Although it normally exists in small amounts, it is very important as a biosynthetic intermediate of phosphoglycerides.

Phosphoglycerides differ from each other with respect to polar head groups (X) whose hydroxyl group is esterified to phosphoric acid (Fig. 4). Since phosphoglycerides contain a polar head group and two non-polar hydrophobic chains of fatty acids, they are amphipathic in nature. The molecule of a phosphoglyceride or, for that matter, of any other phospholipid can be visualised as having a distinct polar and a non-polar domain (Fig. 4), while phosphoglycerides are soluble in most polar organic solvents, they are insoluble in acetone. Figure 5 depicts different polar head groups (X) of common phosphoglycerides.

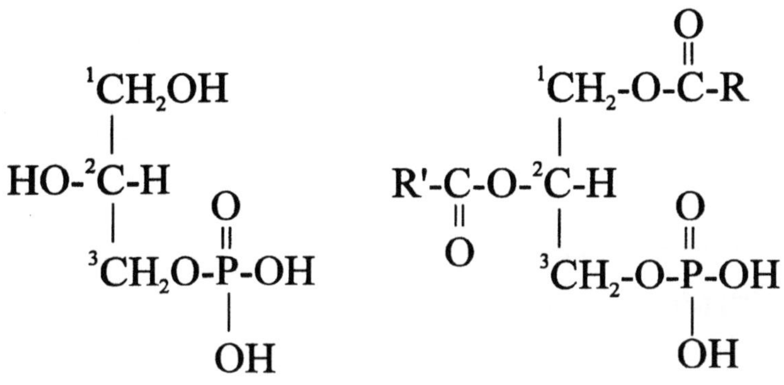

$$^1CH_2OH$$
$$HO-^2C-H$$
$$^3CH_2O-\overset{\overset{O}{\|}}{P}-OH$$
$$OH$$

$$^1CH_2-O-\overset{\overset{O}{\|}}{C}-R$$
$$R'-\overset{\overset{}{\underset{\|}{O}}}{C}-O-^2C-H$$
$$^3CH_2-O-\overset{\overset{O}{\|}}{P}-OH$$
$$OH$$

L-glycerol-3-phosphoric acid   L-phosphatidic acid

Fig. 3. L-glycerol-3-phosphoric acid is the isomer of glycerol phosphate found in natural phosphoglycerides and belongs to the L-stereochemical series. L-phosphatidic acid has two fatty acids, R and R', esterified to the hydroxyl groups at carbon atoms 1 and 2. The parent compound of the series is thus the phosphoric ester of glycerol

Phosphatidylcholine

Phosphatidylcholine (PC) is the most abundant phosphoglyceride in animals and often in higher plants and yeasts. It contains choline as a head group esterified to phosphoric acid (Fig. 5). The complete hydrolysis of PC would thus yield two molecules of fatty acids and one molecule each of glycerol, phosphoric acid and choline. A lyso-derivative of PC, where only one of the two available hydroxyl groups is esterified to a fatty acid, also exists in tissues, albeit in small amounts.

Phosphatidylethanolamine

Phosphatidylethanolamine (PE) is generally the second most abundant phosphoglyceride of eukaryotic cells and occurs as a

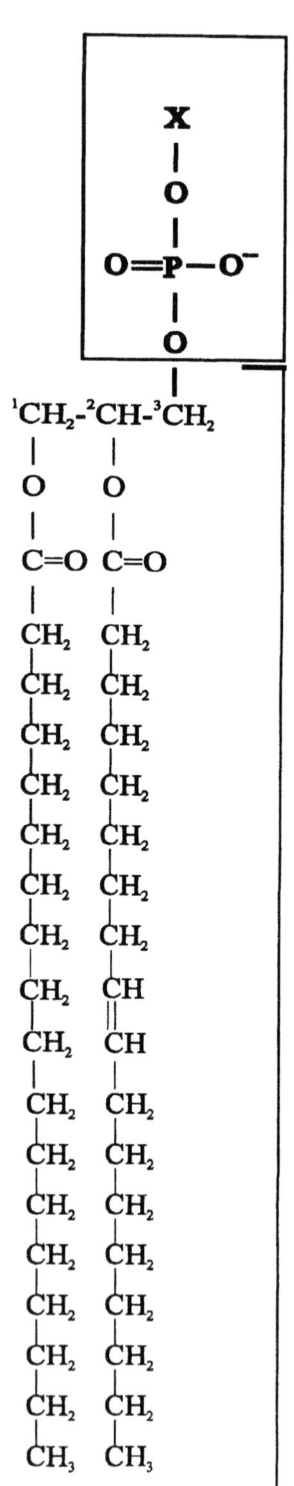

Polar Domain

Non-polar Domain

Fig. 4. Structure of phosphatidic acid. When esterified to X (an alcohol), it represents the general struture of phosphoglycerides. The structure also emphasises the amphipathic nature of phosphogylcerides

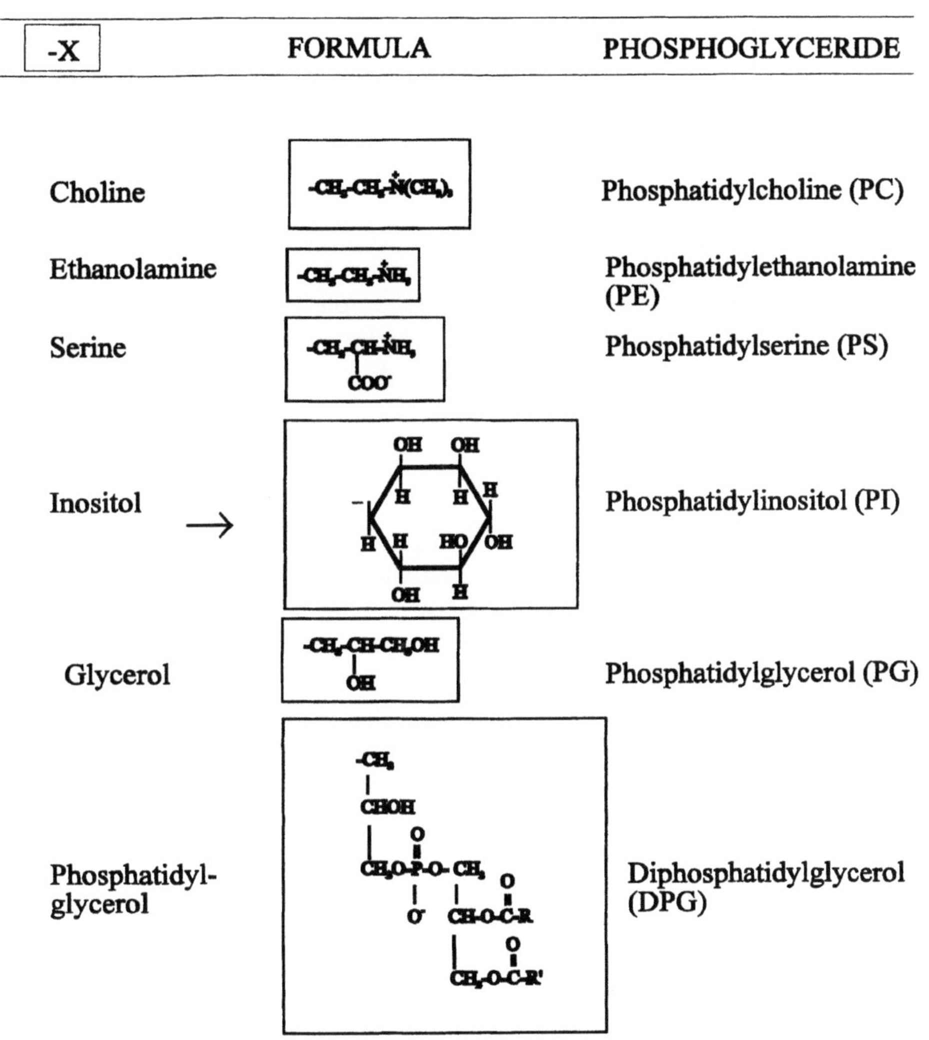

Fig. 5. Head groups of major phosphoglycerides

major phospholipid of bacteria. The complete hydrolysis of PE yields one molecule each of ethanolamine, glycerol, phosphoric acid and two molecules of fatty acids (Fig. 5). The lyso-derivative of PE is rarely found in animal and plant tissues.

Mono- and di-methyl derivatives of PE are also common components of microbial lipids.

Phosphatidylserine

Phosphatidylserine (PS) is a common and predominant component of erythrocyte membrane. It is also found in animal, plant and bacterial cells. The complete hydrolysis of PS would yield one molecule each of serine, phosphoric acid, glycerol and two molecules of fatty acids (Fig. 5).

Phosphatidylinositides

Phosphatidylinositides (PIs) contain a head group of six carbon cyclic sugar alcohol inositol. Phosphatidylinositol (PI) is a common phosphoglyceride of animal, plant and bacterial cells. Similar to other phosphoglycerides, its hydrolysis yields a molecule each of inositol, glycerol, phosphoric acid and two molecules of fatty acids (Fig. 5). The inositol is further phosphorylated to give polyphosphoinositides whose structure, properties and separation strategies are discussed in Chapter VIII, this Volume.

Phosphatidylglycerol and Diphosphatidylglycerol

Phosphatidylglycerol (PG) contains another molecule of glycerol esterified to phosphoric acid and thus its hydrolysis yields two molecules each of fatty acids and glycerol and one molecule of phosphoric acid (Fig. 5). PG is normally present in small quantities in animal tissues. It is a common component of bacterial membranes, where it can also exist as an amino acid derivative, mostly lysine esterified to the 3′ position of the glycerol head group. PG as a derivative of amino acid is often called lipoamino acid.

Diphosphatidylglycerol (DPG), which is commonly known as cardiolipin, is a major phosphoglyceride of mitochondrial membrane lipids, particularly of heart muscle. It consists of a phosphatidylglycerol molecule, whose 3′-hydroxyl group of second glycerol moiety is esterified to the phosphate group of another phosphatidic acid. It can also be visualised as having three molecules of glycerol joined by two phosphodiester bridges; the two hydroxyl groups of external glycerol are esterified with different fatty acids. Thus, its hydrolysis would yield two molecules of phosphoric acid, three molecules of glycerol and four molecules of fatty acids (Fig. 5).

Glycoglycerolipids

These are plant lipids which consist of a diacylglycerol whose 3′-hydroxyl group is linked to a mono- or di-saccharide by a glycosidic linkage. The common sugar which is linked to diacylglycerol is galactose is either mono- and di-galactosyldiacylglycerol, containing one or two molecules of galactose, respectively. Its hydrolysis would yield two molecules of fatty acids, one molecule of glycerol and one or two molecules of galactose (Fig. 6).

Sulfolipids are exclusively found in chloroplast. These consist of monoglycosyldiacylglycerol in which 6th position of the sugar is linked by a carbon-sulphur bond to sulphonic acid (Fig. 6).

4 Sphingolipids

These are complex lipids of brain and nervous tissues as well as important membrane components in both plant and animal cells. This class of lipid does not contain glycerol, instead they have sphingosine (amino alcohol) or its derivative as a

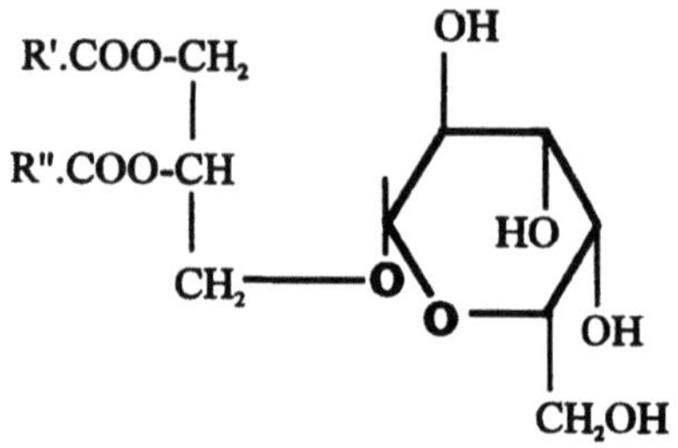

Monogalactosyldiacylglycerol

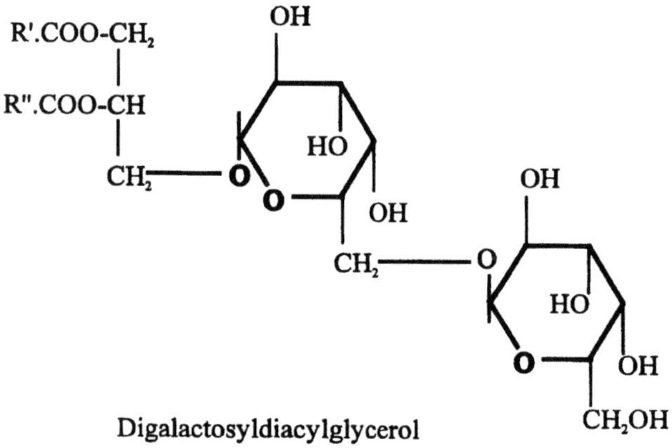

Digalactosyldiacylglycerol

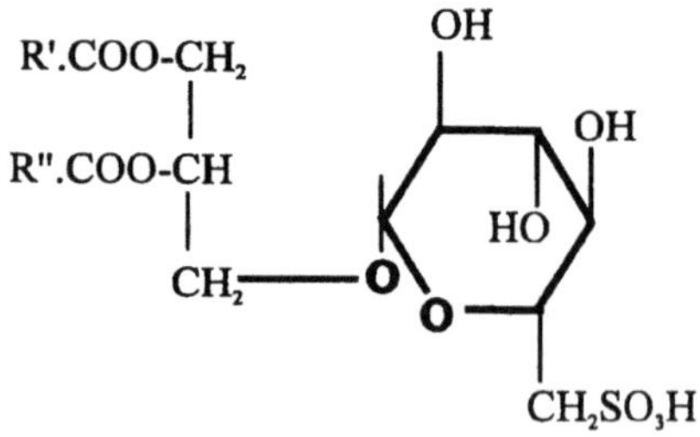

Sulphoquinovosyldiacylglycerol

Fig. 6. Structures of glycosyldiacylglycerols

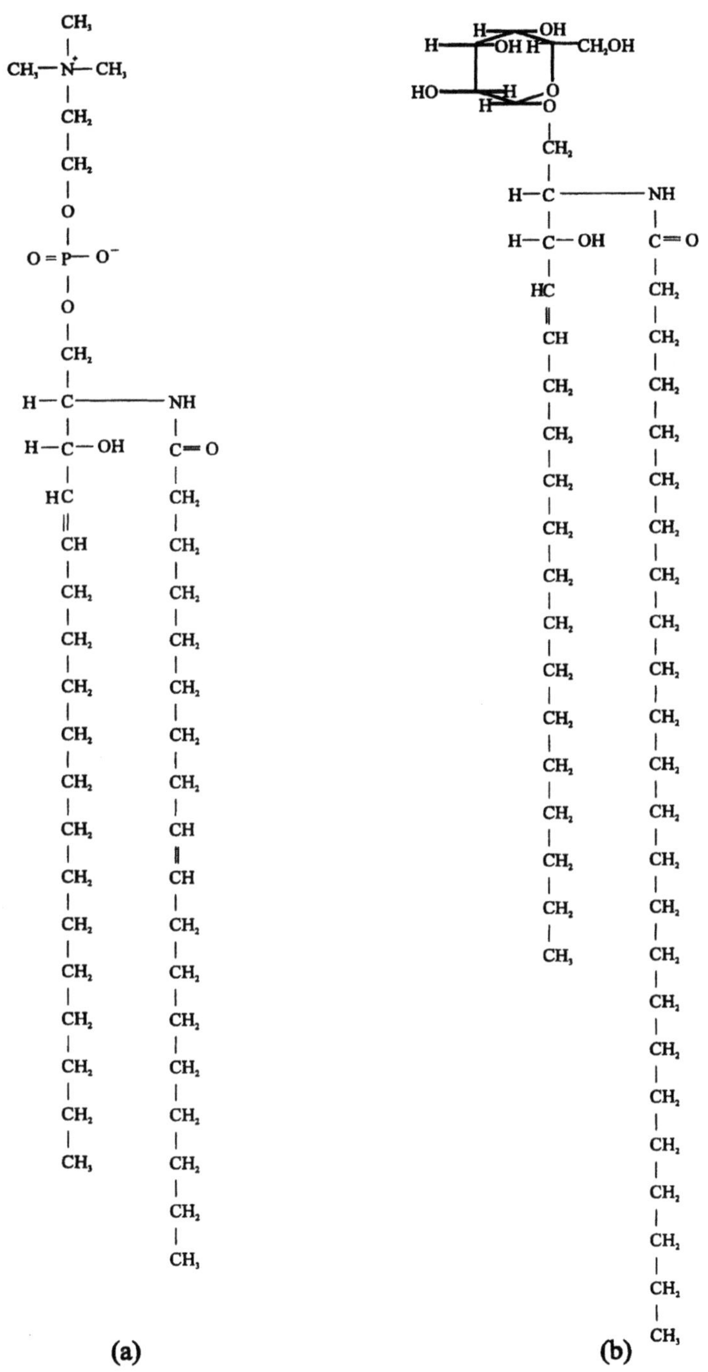

(a)

(b)

Fig. 7. Structure of a representative, **a** sphingomyelin and **b** galactocerebroside

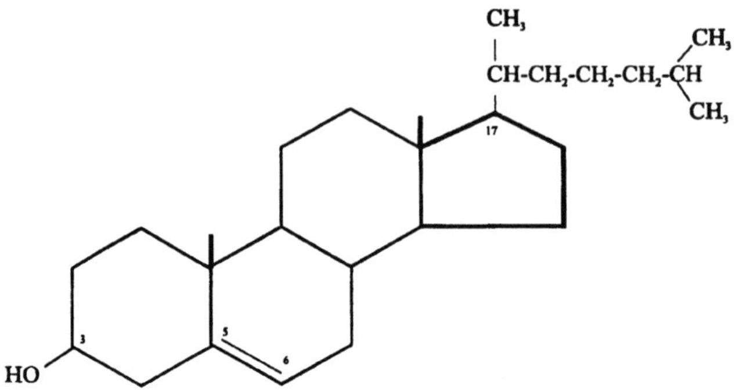

Cholesterol

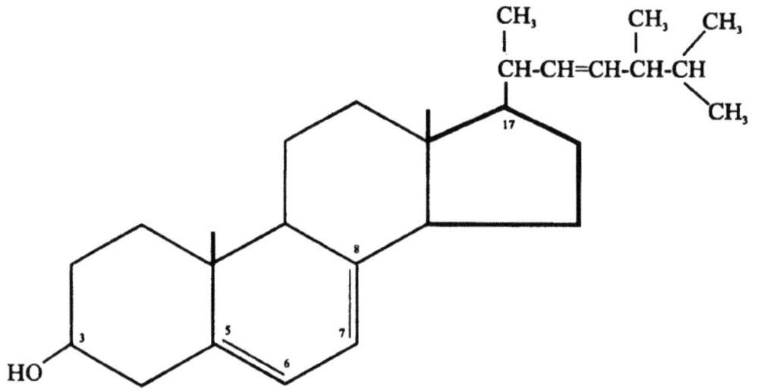

Ergosterol

Fig. 8. Common sterols of animal and fungal membranes.

backbone structure. The amino group of sphingosine is linked via an amide linkage to a long chain fatty acid and the resulting compound is known as ceramide which is a characteristic structure of all sphingolipids.

When phosphorylcholine or phosphorylethanolamine is esterified to the C-1 hydroxyl group of ceramide, the resulting structure is known as sphingomyelin. If the C-1 hydroxyl group is esterified to a sugar as a polar head group, it becomes a cerebroside which could either be gluco- or galactocerebroside, depending upon the nature of head group sugar (Fig. 7).

5   Steroids

Steroids are derivatives of perhydrocyclopentanophenanthrene, a tetracyclic hydrocarbon. While there are variety of steroids identified with many functions from different sources, cholesterol and ergosterol are common steroid alcohols of animal, fungal membranes, respectively (Fig. 8). Bacterial cells do not contain sterols.

Chapters III and IV of this volume describe strategies to isolate, identify and quantitate various lipids which are based on different properties of each class of lipids. The procedure described therein can be adopted to suit the specific requirements of a researcher.

References

Jain MK (1988) Components of biological membranes. In: Introduction to biological membranes, 2nd edn. John Wiley, New York, p 10

Lehninger AL, Nelson DL, Cox MM (1993) Lipids. In: Principles of biochemistry, 2nd edn. Worth Publishers, New York, p 240

Zubay G (1988) Lipids. In: Biochermistry, 2nd edn. Macmillan, New York, p 154

Chapter II   Isolation of Pure Membrane Fractions for Lipid Analysis

L.A. Okorokov[1,2], R. Prasad[3], R.A. Zvyagilskaya[4],
L.P. Lichko[1], T.V. Kulakovskaya[1], N.P. Yurina[4] and
M.S. Odintsova[4]

1   Background

R. Prasad

In order to analyse the lipid composition of different membranes, it is essential that one must be able to isolate pure membranes. As an example, this chapter describes some well-tried protocols for the isolation of different membranes from yeast cells and chloroplast membrane from plant sources. However, there are several protocols available to isolate subcellular membranes from different sources, and the reader is referred to consult those methods for specific requirements (see Methods in Enzymology vol. XXXI Biomembranes Part A; vol. XXXII Part B; Zinser and Daum 1995).

Figure 1 depicts the location of various subcellular organelles in "S-p space". Subcellular particles/organelles can be separated from each other either on the basis of their differences in buoyant density (equilibrium density or isopycnic centrifugation) or on the basis of sedimentation rate (S-value or zonal centrifugation). Normally, a typical protocol of mem-

[1] Institute of Biochemistry and Physiology of Microorganism, Russian Academy of Sciences, Pushchino, Russian Federation
[2] Instituto de Ciêcias Biomédicas, Departamento de Bioquimica Médica, Universidade Federal do Rio de Janeiro, Brasil
[3] School of Life Sciences, Jawahailal Nehrs University, New Delhi 10067, India
[4] Bach Institute of Biochemistry, Russian Academy of Sciences, Moscow-117071, Russian Federation

Table 1. Some (enzymatic) markers for membrane and organelles

Organelle/fragment	Marker (enzyme) activity
Plasma membrane	H^+-ATPase (vanadate sensitive)
Nuclei	DNA
Vacuoles	Bafilomycin A_1 (nitrate) sensitive ATPase, D mannosidase, dipeptidyl aminopeptidase B, proteases A, B, carboxypeptidase Y
Golgi	GDPase, Kex2p (late Golgi)
Endoplasmic reticulum	NADPH-cytochrome "c" oxido-reductase, Dolichyl phosphate mannose synthase, protein mannosyl transferases, Kar2p
Mitochondria	Oligomycin (azide) sensitive ATPase, cytochrome "c" oxydase, cardiolipin
Cytosol	Glucose-6-phosphate dehydrogenase

brane preparation makes use of both methods. Table 1 lists some of the common marker enzymes of various subcellular organelles and their membranes. These enzymes are routinely used to characterise the purity of the membrane preparation. It is not important which method of preparation of a membrane is preferred; what is important, however, is that one must obtain pure membrane fraction for lipid isolation.

2   Isolation of Yeast Plasma membrane by Mechanical Disruption

R. PRASAD

- Grinding medium: 250 mM sucrose; 10 mM Tris-Cl, pH 7.5;   **Solutions**
  1 mM Phenylmethylsulphonyl fluoride (PMSF)
- Suspension buffer: 10 mM Tris-Cl, pH 7.5
- Glass beads (0.45–0.50 mm diameter)
- Percoll

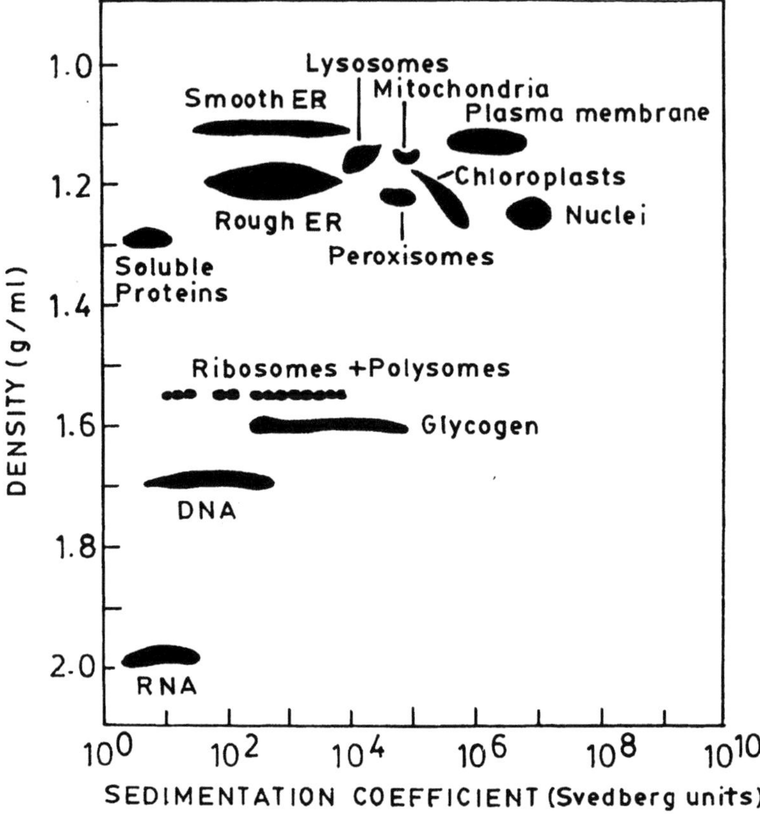

Fig. 1. Subcellular organelles placed in "S-p space". The *abscissa* represent sedimentation coefficient on a logarithmic scale and the ordinate is the equilibrium density. Subcellular organelles can be separated by combination of differential and density gradient centrifugation

Equipment • Braun MSK-Homogeniser, Germany

Preparation of crude membrane (CM)

1. Grow yeast cells in YEPD (YEPD – growth medium containing yeast extract (1%), peptone (2%) and dextrose (2%). Also abbreviated as YPD.) to mid-exponential phase. Harvest the cells by centrifuging at 3000 g for 5 min and wash twice with double distilled water.

2. Suspend the harvested cells (1 g wet weight) in 2 ml of grinding medium and add 2 g of glass beads to this suspension.

3. Mechanically disrupt the cells in a CO_2 refrigerated cell disintegrator (MSK-homogeniser, Germany). Agitate the suspension nine times, each time for 5 s with a gap of 3 s, at 4000 vibrations per minute.

4. Centrifuge the homogenate at 1000 g for 5 min at 4 °C, to remove unbroken cells and glass beads.

5. Wash the pellet once with the grinding medium.

6. Centrifuge the combined supernatants of 4 and 5 steps at 15 000 g for 40 min at 4 °C.

7. Decant the supernatant and resuspend the pellet in suspension buffer.

8. Store the CM, if not required immediately, at −70 °C.

Isolation of plasma membrane (PM)

1. For the preparation of plasma membranes from CM, dilute the CM (25–30 mg protein in 4 ml of suspension buffer) in Percoll (18%, v/v) in 10 mM Tris-Cl, pH 7.5.

2. Centrifuge the suspension at 40 000 g for 40 min in SW-41 Ti type Beckman rotor.

3. The self-generating gradient of Percoll after centrifugation would have the translucent lipid layer on the top and a plasma membrane (PM) band just below that.

4. Aspirate the plasma membrane layer and wash it with 10 mM Tris-Cl, pH 7.5, at 100 000 g for 30 min in the same rotor.

5. Resuspend the plasma membrane pellet in suspension buffer and use directly for lipid extraction. If PM are not required immediately, store at −70 °C till further use.

3   Isolatian of Yeast Plasma Membrane by Spheroplast Formation

L.A. OKOROKOV

! **Note.** There is an opinion that yeast plasma membrane vesicles are leaky to ions and sugars (Serrano 1991) and therefore the fusion of the vesicles with the liposomes is widely used to reduce the passive permeability of membranes (Franzusoff and Cirillo 1983). The following critical points should be taken into account in order to isolate unleaky vesicles with natural lipid composition.

- Spheroplasts should be used as starting material.
- Only low concentrations of DTT (1 mM) should be used during the isolation of spheroplasts. Otherwise the ion permeability of plasma membranes increases and membranes are deenergised (Petrov and Okorokov 1992).
- H^+-ATPase is activated the sheroplast incubation with glucose.
- Use of protease inhibitors is essential during all steps of isolation of membranes.

Solutions
- Solution A: 0.7 M sorbitol in 25 mM Tris-Cl buffer (pH 7.2) containing 1 mM dithiothreitol (DTT)
- Solution B: 0.8 M sorbitol in 25 mM Tris-Cl (pH 7.2) containing 1 mM PMSF (250 mM PMSF in ethanol should be diluted immediately before the use of solution)
- Solution C: 0.7 M sorbitol in 25 mM MOPS-NaOH or MES-NaOH (pH 6.5), containing 100 mM glucose
- 25 mM Tris-EDTA, pH 7.2 in 0.7 M sorbitol
- 250 mM PMSF in ethanol (keep in freezer and warm before use)
- Solution of protease inhibitors: antipain, aprotinin, chymostatin, pepstatin, leupeptin (1 mg/ml, each; keep in freezer and warm in ice before use). Aliquot this solution for storage

The following protocol is the modification of our published **Procedure**
method (Okorokov and Petrov 1986).

1. Grow yeast cells in YPD to mid-log phase of growth and
   collect them by centrifuging at $3000\,g$ for 5 min.

2. Suspend the cells with distilled water at room temperature
   and combine suspensions in one tube of determined
   weight. Repeat the centrifugation, remove the supernatant
   and determine the weight of wet cells.

3. To convert the cells to spheroplasts, resuspend them in
   solution A, containing the snail lytic enzyme or lyticase.
   The typical ratio is 50 mg of snail enzyme in 5 ml of solution A containing 1 g wet cells. Incubate the cell suspension
   with gentle shaking at 30 °C.

4. Take 10 µl sample of the suspension at the beginning
   of the incubation, add 990 µl water, mix and read A_{600} after
   1 min against water.

5. Monitor the formation of spheroplasts by the decrease
   of A_{600} during 15–60 min owing to osmotic lysis of
   spheroplasts in water. Incubation for 45–60 min normally
   gives the decrease of about 70–90% of the initial A_{600}
   value.

6. Add 250 mM solution of PMSF (in ethanol) to a final concentration of 1 mM and 25 mM EDTA in solution A
   to a final concentration of 1 mM, cool the spheroplast suspension on ice.

Note. It is very important to make all the following operations   **!**
on ice. The solutions, centrifuge tubes, rotors and homo-
genisers should be precooled.

7. Put cold solution B in appropriate tubes for SS-34 rotor
   (15–20 ml in each tube) and overlay carefully with
   10–18 ml of spheroplast suspension. Use inclined posi-

tion of tubes and slow flux of the suspension during overlaying.

8. Centrifuge at 1100 g for 5 min (SS-34 of Sorvall or rotor 7 of Hitachi CR-21). Remove the supernatant carefully, vortex the pellet gently with 2 ml of solution C and combine the suspensions of spheroplasts.

9. Incubate the spheroplast suspension in a water bath for 10 min at 30 °C with gentle stirring.

! **Note.** This incubation makes it possible to activate H^+-ATPase of plasma membrane (Serrano 1983) and to open some spheroplasts owing to metabolic lysis (Indge 1968). To prevent the metabolic lysis of spheroplasts add $MgSO_4$ to a final concentration of 40 mM. Since the role of lipids in the regulation of the H^+-ATPase by glucose is not investigated yet (Serrano 1991) it might be interesting to compare the lipid composition of plasma membranes before and after the glucose activation of ATPase.

10. Cool the suspension on ice and then add 2 volumes of cold solution of 25 mM Tris-EDTA, pH 7.2.

11. Add a mixture of protease inhibitors (10 µl for each 10 ml of the suspension) and PMSF immediately before the homogenisation (final concentration of PMSF should be 1 mM).

12. Make 15–20 full strokes in Dounce homogeniser.

13. Centrifuge the supernatant at 8000 g for 20 min to obtain a crude membrane (CM) fraction, enriched with plasma membranes (PM) and mitochondria (MIT).

14. Discard the supernatant, vortex the pellet and add quickly 10–15 ml of Tris-acetate, pH 5.0.

15. Vortex again and centrifuge at 500 g for 2 min.

Note. This step makes it possible to open the plasma membrane vesicles and remove the cytoplasmic fragments which might have trapped during the spheroplast lysis. The acidification of the suspension (pH 5) allows mitochondria to aggregate and thus they are separated from plasma membranes by centrifugation (Fuhrmann et al. 1976; Dufour and Goffeau 1978).

16. If some aggregated mitochondria were found floating, filter the supernatant through the glass fibre paper (Whatman or Sartorius).

17. Add 1 N NaOH or dry Tris base to the supernatant obtained from step 15 while stirring to bring the suspension to pH 7.2.

Note. It is essential to go through the "acidification steps" as soon as possible, within 10–15 min. The acidification inactivates H^+-ATPase of plasma membranes (Dufour and Goffeau 1978), which may be an important consideration if one wants to use PM preparation for ATPase activity.

18. Sediment the PM sheets at 20 000 g for 10 min and wash the pellet with 15–20 ml of 25 mM Tris-Cl or MES-NaOH, pH 7.2. Recentrifuge at 20 000 g for 10 min.

19. Add 2–3 ml of 25 ml MOPS-Na pH 7.2 to the pellet, vortex and add quickly during vortexing the equal volume of 0.5 M sucrose in 25 mM MOPS-Na pH 7.2.

20. Sediment the plasma membrane vesicles at 20 000 g for 10 min, resuspend pellet in the same buffer, aliquot and store for the lipid analysis or for transport investigation.

4 Isolation of Con A-Modified Plasma Membranes

Mannoproteins of the extracellular surface of the yeast plasma membranes can interact with plant lectin Concanavalin A (Con A), isolated from soyabeans, if the spheroplasts are incubated with the lectin before their lysis (Duran et al. 1975). The subsequent lysis of the Con A-modified spheroplasts results in the formation of heavy plasma membrane sheets, which can be pelleted by low speed centrifugation. Different organelles and vesicles, derived from intracellular membranes, will stay in the supernatant (Scarborough 1988; Okorokov 1994).

Procedure

1. Prepare the spheroplasts according to the procedure described in Section 3 and centrifuge their suspensions through the layers of 20 ml of solution B at 1100 g for 5 min.

2. Vortex the pellet gently, add about 2 ml of solution A to each tube and vortex again.

3. Combine the suspensions of spheroplasts in one tube for SS-34 rotor and suspend the rest of the pellets with a soft paint brush in solution A.

4. Centrifuge at 1100 g for 5 min, remove the supernatant and vortex the pellet gently.

5. Add 10 ml of solution A, containing 2 mM $MgCl_2$ and carefully suspend the spheroplasts and the rest of the cells with a paint brush.

6. Add solution of Con A (25 mg/ml of solution A, containing 1 mM $MnCl_2$ and 1 mM $CaCl_2$, pH 7.2) to give a ratio of 3.5 mg Còn A per g wet weight of the cells used in preparing spheroplasts.

7. After 15–17 min incubation at 0 °C, sediment the spheroplasts at 120 g for 10 min, wash once with 20 ml of

solution A and sediment at 120 *g* for 10 min. The pellet of Con A-treated spheroplasts is ready for lysis. The pellet is resuspended in lysis buffer and processed as outlined in steps 6 to 9 of Section 3. The PM fraction obtained by the Con A method does not contain significant cross-contamination of endoplasmic reticulum, mitochondria, Golgi and vacuoles.

5 Isolation of Intracellular Organelles of Yeast and Their Membranes

L.A. OKOROKOV

The isolation of the yeast intracellular membranes such as endoplasmic reticulum, Golgi, secretory vesicles etc. is a difficult task. The more acceptable approach is to use temperature-sensitive mutants of the protein secretory pathway (Novick et al. 1980). Accumulation of vesicles of ER or Golgi or secretory vesicles due to the corresponding mutation makes it possible to enrich the intracellular membranes of the organelle of choice. This good starting object needs, however, appropriate techniques for the separation of the desirable membranes from other membranes. The following protocol is an example of the separation of the intracellular membranes from a wild-type yeast strain of *S. cerevisiae* ×2180. This protocol can be adapted for any secretory mutant when the spheroplasts are available.

Reagents

- Solution A: 1.2 M sorbitol in 50 mM Tris-Cl, pH 7.2; containing 40 mM β-mercaptoethanol
- Solution B: 1.4 M sorbitol in 50 mM Tris-Cl, pH 7.2
- Solution C: 365 mM sucrose, 20 mM MOPS-Na (pH 7.2), 1 mM DTT
- Sucrose solutions: 56, 52, 48, 45, 42, 39, 36 and 33% (w/v) in 10 mM HEPES-Na, pH 7.2

Procedure

1. Prepare the spheroplasts at 30 °C with solution A. For 1 g of wet cells (mid-log phase yeast cells), use 5 ml of solution A and 50 mg of lyophilised snail lytic enzyme (with lyticase from Sigma, use their recommendations).

2. Incubate the cell suspension with gentle shaking.

3. The formation of spheroplast could be monitored by decrease in A_{600} as described in Section 3, step 4–7.

4. Overlay 10–18 ml of suspension of spheroplast on 20 ml of solution B.

5. Centrifuge at 1100 g for 5 min in a SS-34 rotor.

6. Wash the pellet with 20 ml of solution A.

7. Prepare the lysate with lysis solution C and remove Con A modified PM (Section 4). Centrifuge the suspension of intracellular membranes at 35 000 g for 15 min and remove mitochondria as described in Section 3, step 14–20.

! **Note.** It is important to completely remove Con A, otherwise it could modify the mannoproteins of intracellular membranes.

8. Add 2 ml of solution C to each pellet and vortex gently; combine the suspensions of spheroplasts.

9. Incubate the spheroplast suspension at 30 °C for 10 min with gentle stirring.

10. Cool the suspension on ice and add 3 volumes of 25 mM Tris-Cl pH 7.2 without EDTA. (see recommendations of steps 11, 12, Secto 3). Centrifuge the lysate at 120 g for 10 min and save the supernatant.

11. Centrifuge it at 590 g for 5 min. Vortex the white pellet of ConA-modified plasma membranes, add 5 ml of 25 mM Tris-Cl pH 7.2 in 0.2 M sorbitol, vortex and centrifuge at 590 g for 5 min.

12. To the pellet add 1–2 ml of the same buffer, vortex and make 5 strokes in a small Dounce homogeniser. Aliquot the suspension and store it at $-70\,°C$ or in liquid nitrogen.

6   Fractionation of Membranes on a Sucrose Density Gradient

L.A. Okorokov

1. To prepare sucrose gradient, overlay (in the appropriate tube for a SW-41 rotor) the sucrose solutions, starting with 1.33 ml of 56% and following with 52%–39% (each of 1.33 ml). Overlay then 1 ml of 36% sucrose, 0.7 ml of 33% and 0.7 ml of 30% sucrose.

2. Each layer should receive 2 µl of the set of protease inhibitors.

3. Load on the top of the gradient about 1 ml of the membrane suspension.

4. Centrifuge at 174 000 g for 2 h and 45 min and fractionate the gradient in a cold room (4 °C).

5. Assay for different marker enzymes will show the localisation of different membrane fractions in the gradient (Table 1 Okorokov 1994; Zinser and Daum 1995). After ascertaining the purity of the desired subcellular organelles or their membranes, these membrane fractions can be used to extract and quantify the lipids.

Note. Fig. 2. shows the example of such fractionation when ATP-dependent and protonophore-insensitive Ca^{2+} uptake was measured for each fraction. Peaks III and IV contain membranes which form two different subcompartments of rough endoplasmic reticulum, while peaks V and VI represent two subcompartments of smooth endoplasmic reticulum. Golgi membranes migrate in fractions 22–23, while vacuoles are in the peak VIII (markers are not shown).

All eight peaks (compartments) have different Ca^{2+}-ATPases of ER type (about 4). Peaks III and VI are probably characterised by the presence of the same Ca^{2+}-ATPase, while peaks IV and V have another one. Third Ca^{2+}-ATPase is probably in peak II and VIII, fourth one is in peak VII.

7   Isolation of Mitochondria of Yeast

R.A. Zvyagilskaya

Solutions
- Buffer 1:0.05 M Tris-Cl, 10 mM DTT, pH 8.0–8.3.
- Buffer 2:0.01 M Citrate-phosphate, 1 M xylitol, 50 mM EDTA, pH 5.8.
- Buffer 3:0.01 M Tris-Cl, 0.4 M mannitol, 1 mM EDTA, 0.4% bovine serum albumin (BSA, fatty acid free), pH 7.4.
- Buffer 4:0.01 M Tris-Cl, 0.6 M mannitol, 1 mM EDTA, 0.4% bovine serum albumin (BSA, fatty acid free), pH 7.0.
- 1 M xylitol
- Snail gut juice from *Helix pomatia*

Procedure
1. Grow yeast cells to late exponential growth phase in a semi-synthetic media containing glycerol as the sole carbon source. Harvest the cells and wash twice in ice-cold sterile distilled water.

2. Suspend the washed cells (15 g wet wt) in 150 ml of Buffer 1 and incubate at room temperature for 30 min, then collect the cells by centrifugation at 2500 g for 7 min.

3. Wash the cells twice with ice-cold distilled water.

Note. This step is necessary to remove DTT completely.

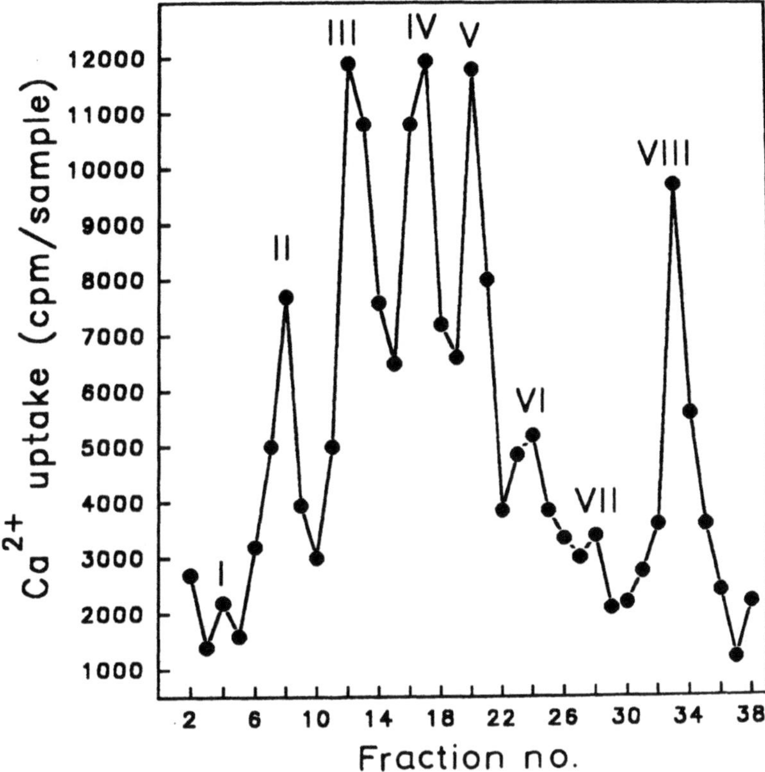

Fig. 2. $^{45}Ca^{2+}$-uptake by different fractions of yeast membranes after their fractionation in the sucrose density gradient

4. Resuspend pre-treated cells in 150 ml of spheroplasting buffer 2. To this suspension, add 15 mg of Helicase per g of cells (wet wt).

5. Incubate the yeast suspension at 30 °C with constant stirring for their conversion to spheroplasts. The spheroplasts formation can be monitored by checking the lysis of the suspension (as described in Sect. 3).

6. After 20 min (the absorbance will drop about fourfold), collect the spheroplasts by centrifugation at 3000 g for 10 min and wash twice in 75–100 ml of 1 M xylitol.

7. Resuspend the pellet in 15 ml of buffer 3 (chilled on ice) and homogenise by 20 strokes in a tight-fitting Dounce homogeniser.

8. Dilute the homogenate immediately with 9 volumes of ice-cold buffer 4 and centrifuge at 1200 g for 10 min.

9. Centrifuge the resulting supernatant at 6500 g for 20 min.

10. Resuspend the mitochondrial pellet in 0.2 ml of buffer 4 (to give 20–25 mg protein/ml). The yield of mitochondrial protein should be 30–35 mg per 15 g of cells (wet wt).

Note: This protocol provides a mitochrondria, preparation useful for functional biochemical analysis. For a PM-fre mitochondria preparation remove ConA-modified PM according to Section 4 and sediment mitochondria at 6500 g for 20 min.

11. Keep the suspension of mitochondria on ice and use for extraction of lipids (as outlined in Chap. III, this Vol.).

8   Isolation of Vacuoles of Yeast

L.P. Lichko, T.V. Kulakovskaya and L.A. Okorokov

Solutions
- Solution A: 0.7 M mannitol, 10 mM DTT, 0.14 M Sodium citrate (pH 6.7)
- Solution B: 0.1 M sorbitol, 10 mM sodium citrate (pH 6.8)
- Solution C: 10 mM Tris-Cl (pH 7.5), 1 mM EDTA, 2 mM DTT, 0.5 mM PMSF

Isolation of spheroplasts
1. Harvest yeast cells grown in YEPD to the mid-exponential phase, by centrifuging at 3000 g for 5 min.

2. Suspend the cells with distilled water at room temperature and combine them in a preweighed tube. Repeat centrifugation, remove the supernatant and detenmine the weight of wet cells.

3. To convert the cells to spheroplasts, resuspend them in solution A containing the snail lytic enzyme or lyticase. For 30 g wet cells add 240 ml of solution A containing 3.6 g of lyophilised snail gut juice.

4. Incubate the cell suspension with shaking at 30 °C. The formation of spheroplasts is monitored by following the decrease in A_{600} as described in Section 3, step 4, 5.

5. Centrifuge the suspension of spheroplasts, cells and rest of cell walls at 1100 g for 5 min. Remove the supernatant carefully, vortex the pellet gently, add in each tube 3–5 ml of solution A and transfer the suspension in one tube appropriate for an SS-34 rotor. Centrifuge at 1100 g for 5 min and save the pellet.

Lysis of spheroplasts

1. Suspend the pellet in 55 ml of 4% Ficoll, prepared in solution B supplemented with 1 mM PMSF, directly before the lysis of spheroplasts.

2. Homogenise the spheroplasts with 20 full strokes and transfer 15 ml of lysate in each tube appropriate for a Beckman SW-27 rotor.

Isolation of vacuoles

3. Overlay a discontinuous gradient with solution B containing 2% Ficoll (15 ml) and solution B (8 ml). This should be done carefully in order to prevent a mixing of layers. Use inclined position of tubes (about 45°–60°) and slow flux of solutions during overlaying. Centrifuge the content at 25 000 rpm for 30 min (Beckman, SW-27 rotor).

4. After centrifugation, vacuoles form a distinct band between solution B containing 2% Ficoll and solution B. The lipid

particles separated from vacuoles will be localised on the top of the gradient.

5. Remove carefully the vacuolar band with a syringe equipped with a wide needle. One could use a steel needle which has a "J-like" end. This would make it possible to collect the vacuolar layer from below. Dilute the vacuolar suspension four times with solution B and sediment the vacuoles by centrifuging the suspension at 35000 g for 50 min.

6. Remove the supernatant, vortex the pellet gently, add 200 µl of solution B and vortex again. Combine this vacuolar suspension with suspensions prepared identically from other tubes and homogenise the total suspension for five strokes, aliquot and store vacuoles at −70 °C or use immediately.

Isolation of vacuolar membranes

1. Subject vacuoles to osmotic lysis by suspending the pellet in 5 ml of solution C.

2. Separate the membrane fraction from soluble vacuolar proteins by centrifugation at 30000 g for 20 min.

3. Wash the vacuolar membranes twice with the same buffer.

4. The isolated vacuolar membranes contain about 70% proteins, 85% α-mannosidase activity and 90% vacuolar ATPase activity of intact vacuoles (Okorokov et al. 1982; Roberts et al. 1991).

5. Use these membranes directly for lipid extraction as per the protocols discussed in Chapter III, this Volume.

9 Chloroplast Isolation

N.P. YURINA AND M.S. ODINTSOVA

The methods of chloroplast isolation usually involve brief grinding of a suitable leaf tissue in ice-cold buffer containing osmoticum in an efficient homogeniser followed by rapid centrifugation to separate the organelles from damaging components in the homogenate. To speed up the preparation, the preliminary centrifugation (to remove nuclei and "cell debris") is usually replaced by filtration through several layers of muslin. Sometimes the homogenate is subjected to two cycles of differential centrifugation: a low speed spin (e.g. 500 g for 2 min) to remove nuclei and "cell-debris", followed by a higher-speed spin (e.g. 6000 g for 15 min) to pellet the actual chloroplast preparation.

The isolated chloroplast preparation still contains some broken chloroplasts. The quality of chloroplasts may be further improved by resuspending the organelles in homogenising medium and rapid pelleting for a second time, when some of the damaged chloroplasts remain in the supernatant. The specific details of the procedures for chloroplast isolation should be developed for each plant species and often depend on the goal of the experiment (Palmer 1986).

As an example, the procedure of chloroplast isolation from 10–12-day-old pea seedlings is described below (Yurina et al. 1991).

- Isolation medium:
  - 330 mM mannitol,
  - 50 mM Tris-Cl (pH 8.0),
  - 5 mM EDTA,
  - 10 mM dithiothreitol (DTT) and
  - 0.1% bovine serum albumin (BSA).

Solutions

Note. DTT (10 mM) and BSA (0.1%) are added to the buffer immediately before chloroplast isolation.

1. Mince shaded (1- to 3-day) pea seedlings in knife-blade homogeniser at 10 000 rpm for 10 to 15 s in the isolation medium. The tissue:buffer ratio should be 1:3 (w/v).

Isolation of chloroplasts

2. Remove the cell debris and nuclei by successive filtration of the homogenate through 20- and 50-mesh nylon sieves without applying pressure.

3. Centrifuge the filtered homogenate at 400 g for 5 min and resuspend the pelleted chloroplast in the isolation medium.

4. Centrifuge it for the second time at 1500 g for 15 min.

5. After the second centrifugation, resuspend the chloroplast pellet in 10 volumes of isolation medium.

Purification of chloroplast

1. Apply 10 ml of the suspension to a discontinuous sucrose gradient prepared from 52% (30 ml) and 30% (15 ml) solutions of sucrose in 50 mM Tris-Cl buffer, pH 8.0, containing 20 mM EDTA. Centrifuge the samples at 25 000 rpm for 2 h in a Beckman centrifuge with an SW-28 rotor. Collect the green zone separating the 30 and 52% sucrose layers.

2. Dilute the fraction collected from the gradient centrifugation with three volumes of the isolation medium and centrifuge at 3000 g for 10 min. Use the sedimented pellet of purified chloroplast directly for lipid extraction.

   All the steps for isolation and purification of chloroplasts should be carried out at 4 °C.

In addition to purification of chloroplasts using a discontinuous sucrose gradient, a linear (40–60%) Percoll gradient can also be used. Preparation of Percoll gradient and dilution of concentrated Percoll solution is done using 0.25 M sucrose in 50 mM Tris-Cl buffer, pH 8.0. The samples are centrifuged at 2500 rpm for 45 min. Lower green zone containing chloroplast is collected using a syringe with a wide-tip needle.

The chloroplast fraction is diluted 1:3 with the buffer containing 50 mM Tris-Cl, pH 8.0, 20 mM EDTA and pelleted by centrifugation at 3000 g for 10 min. The purified chloroplast fraction contains 90% or more unbroken chloroplasts.

The analysis of purified chloroplast by analytical CsCl-density centrifugation and digestion by restriction endonucleases of DNA from chloroplast fraction shows the absence of mitochondrial and nuclear contamination. Rates of photosynthetic reduction of carbon dioxide and associated oxygen evolution can also be used as indicators of the functional competence of the chloroplast preparations. The purity of the chloroplast fraction can be checked by the analysis of marker enzyme. The purified chloroplast prepared from either gradient can be used directly for lipid extraction.

References

Dufourc JP, Goffeau A (1978) Solubilization by lysolecithin and purification of the plasma membrane ATPase of the yeast *Schizosaccharomyces pombe*. J Biol Chem 253:7026–7027

Duran A, Bowers B, Cabib E (1975) Chitin synthetase zymogen is attached to the plasma membrane. Proc Natl Acad Sci USA 72:3952–3955

Franzusoff A and Cirillo VP (1983) Glucose transport activity in isolated plasma membrane vesicles from Saccharomyces cerevisiae. J Biol Chem 258:3608–3615

Fuhrmann GF, Boehm C, Theuvenet APR (1976) Sugar transport and potassium permeability in yeast plasma membrane vesicles. Biochim Biophys Acta 433:583–596

Indge KI (1968) Metabolic lysis of yeast protoplasts. J Gen Microbiol 51:433–440

Novick P, Field C, Schekman R (1980) Identification of 23 complementation groups required for post-translational events in the yeast secretory pathway. Cell 25:451–460

Okorokov LA (1994) Several compartments of *Saccharomyces cerevisiae* are equipped with Ca^{2+}-ATPase(s). FEMS Microbiol Lett 117:311–318

Okorokov LA, Petrov VV (1986) Isolation of plasma membrane vesicles from the yeast *Saccharomyces carlsbergensis* suitable for solute transport studies. Biol Membr (USSR) 3:549–556

Okorokov LA, Kulakovskaya TV, Kulaev IS (1982) Solubilization and partial purification of vacuolar ATPase of yeast *Saccharomyces carlsbergensis*. FEBS Lett 145:160–162

Palmer JD (1986) Isolation and structural analysis of chloroplast DNA. Methods Enzymol 118:167–186

Petrov and Okorokov (1992) Mercaptoethanol and Dithiothreitol decrease the difference of electrochemical proton potentials across the yeast plasma and vacuolar membranes and activate their H^+-ATPases. Yeast 8:589–598

Roberts CJ, Raymond CK, Yamashiro CT, Stevens TH (1991) Methods for studying the yeast vacuoles. Methods Enzymol 194:644–661

Scarborough G (1988) Large scale purification of plasma membrane H^+-ATPase from a cell wall-less mutant of *Neurospora crassa*. Methods Enzymol 157:574–579

Serrano R (1983) In vivo glucose activation of the yeast plasma membrane ATPase. FEBS Lett 156:11–14

Serrano R (1991) Transport across yeast vacuolar and plasma membranes. The molecular and cellular biology of the yeast *Saccharomyces*: genome dynamics, protein synthesis and energetics. I:523–585. Cold Spring Harbor Laboratory Press

Yurina NP, Belkina GG, Piletskaya, Karapetyan NV, Odintsova MS (1991) Isolation and characterization of chloroplast DNA of *Azolla pinnata*. Prikl Biokhim Mikrobiol (Russian) 29:299–307

Zinser E and Daum G (1995) Isolation and biochemical characterization of organelles from yeast, *Saccharomyces cerevisiae* Yeast 11:493–536

ANJNI KOUL AND RAJENDRA PRASAD

1   Background

Lipids form a group of organic compounds which is widely distributed in living systems. The term lipid cannot be defined concisely, but it refers to a large number of compounds which have similar solubility properties (for more details on type of membrane lipids, the reader is referred to Chap. I, this Vol.). Membrane lipids are amphipathic molecules which are water-insoluble and can be generally extracted from different sources by treatment with the combination of polar and non-polar organic solvents. The aim of the present chapter is to consider practical aspects of isolation of lipids. The chapter is intended to provide the reader with necessary information for the extraction of membrane lipids with confidence. The extraction protocols described herein should provide the reader with sufficient background to adopt the procedure to fit to the experimental requirements.

For the purpose of extraction, membrane lipids can be classified into three distinct groups:

Non-Polar Lipids

This group includes carotenoids, hydrocarbons, sterol esters, wax esters and glycerides. These lipids are often extractable

School of Life Sciences, Jawaharlal Nehru University, New Delhi-110067, India with ether or chloroform, since they are weakly bound to each other and to the hydrophobic regions of the membrane by Van der Waals forces.

Polar Lipids

This group includes phosphoglycerides, glycolipids and sterols which are not only held by Van der Waals forces but are also bound by hydrogen bonds and, in some cases, by electrostatic interactions. These bonds are less readily disrupted and, therefore, lower alcohols or acetone must be used to free these lipids from membranes. Chloroform:methanol (2:1, v/v) is commonly used as an extraction mixture which removes polar lipids from membranes. However, acidic phospholipids and polyphosphoinositides are not generally extractable by this combination of polar and non-polar solvents, one must consider adding 0.25% HCl (conc.) to the solvent mixture for complete extraction of these lipids. Since plasmalogens would be hydrolysed by the acid, the use of acid should be avoided if plasmalogens are under investigation (see Chap. VII, this Vol. for more details).

Anchored Lipids

This group includes fatty acids and hydroxy fatty acids bound covalently through ester, amide or glycosidic bonds, predominantly to polysaccharide structures. These fatty acids are extractable only after acidic or basic hydrolysis of the covalent bonds.

Before the extraction of membrane lipids, it is imperative to know that one is dealing with desired pure membrane fraction (discussed in Chap. II, this Vol.). Membrane lipids from animal, plant and microorganism sources should ideally be extracted as soon as possible after their preparation to avoid changes in lipid components. When this is not possible, the membrane fraction should be frozen rapidly on dry ice and stored in a sealed glass container at $-70\,^{\circ}C$ in an atmosphere of nitrogen. Under these conditions, the sample can be kept for several months.

2   Extraction Protocols of Membrane Lipids from Different Sources

General Principles of Extraction. Folch and Stanley (1957) procedure of lipid extraction is the most common method used. However, modifications to this method have also been adopted. Twenty volumes of chloroform:methanol (2:1, v/v) mixture per volume of tissue or membrane pellet or suspension is recommended to yield a single homogeneous suspension. If the ratio is small (12–15 volumes) two phases will be formed, more solvent would be required to obtain a single phase. The single phase of solvent provides better interaction of polar and non-polar solvents with membrane lipids. High ratio is probably harmless to some extent, but too high a ratio of solvent will lower the water content, making polar lipid extraction incomplete.

There are several protocols of lipid extraction which differ slightly from each other. Membrane lipid extraction protocols commonly used for membranes derived from different sources are described below. These protocols can also be used for the extraction of total lipids from tissues of different origin.

- Benzene
- Chloroform
- Methanol
- Propan-2-ol
- $CaCl_2$ anhydrous

Reagents and solvents

- 0.9% NaCl
- Nitrogen gas

Equipment • Thermal block or rotary evaporator

Extraction from red blood cells 1. Resuspend the RBC membrane fraction pellet directly in glass tubes (with stoppers) in at least 20 volumes of chloroform:methanol (2:1, v/v) or more commonly used alcohol:ether (3:1, v/v). Shake the suspension vigorously for 2–3 min. (When using tissue for lipid extraction, it is necessary to homogenise the tissue extract in a glass homogeniser for uniform suspension.)

! **Note.** Normally 20 volumes of solvent mixture is sufficient to make a single homogeneous phase. If the suspension does not form a single phase, more solvent mixture can be added.

2. Flush the suspension with N_2 gas and tightly close with stoppers. The tubes can be shaken gently in a shaker at room temperature for 15–30 min.

3. Separate the denatured protein either by filtration or centrifugation. We recommend centrifugation where denatured proteins can be removed by discarding the pellet.

4. Transfer the supernatant to another clean and dry graduated centrifuge tube and mix with 0.2 volumes of 0.9% NaCl (to remove non-lipid contaminants). The resulting turbid suspension could separate into two phases within ca. 2 h. However, a quick centrifugation in any bench top centrifuge will separate the two phases within a few minutes.

5. Remove the upper layer containing non-lipid contaminants carefully with a Pasteur pipette or aspirate off using a suitable device.

6. Recover the lower clear phase containing most of the lipids in another tube and concentrate as described below (for complete extraction of membrane lipids, e.g. for the extraction of polyphosphoinositides, lysophospholipids and gangliosides special steps are necessary; see Chap. VIII, this Vol.). For quantitative extraction, insoluble residue of step 3 can be re-extracted with fresh solvent.

The method of lipid extraction is essentially similar to animal tissues. However, in order to inactivate hydrolytic enzymes, e.g. lipases, which are activated in the presence of chloroform and diethyl ether, it is necesssary to boil the plant tissue or its membrane fraction in propan-2-ol before chloroform: methanol extraction (Nichols 1963). In addition, for quantitative extraction of polar lipids, it might be necessary to use high salt buffer to partition the solvent phase. A typical protocol for the extraction of plant lipids is described below:

Extraction from plant tissues

1. Boil the plant tissue or protoplast or membrane fraction prepared from it in five volumes of propan-2-ol (v/v) for 5 min.

2. Macerise the boiled fraction in a mortar in the presence of washed inactive sand and propan-2-ol.

3. Transfer the extract to a glass stoppered tube and add 20 volumes of chloroform:methanol. Flush the tube with N_2 and shake the mixture for 5 min.

4. Filter the total extract through Whatman # 1 filter disc. For quantitative extraction, the residue on the filter paper can be re-extracted with fresh solvent mixture.

5. Wash the filtered extract with 0.2 volumes of NaCl (0.9%) to remove non-lipid contaminants and transfer the extract to a separating funnel to separate aqueous and non-aqueous phases. After flushing with N_2, leave the funnel overnight for the separation of two phases.

6. Collect carefully the lower phase from the separating funnel which contains most of the lipids in a stoppered glass tube. Flush N$_2$ and store at 4°C for evaporation and concentration.

Extraction from yeast

This protocol is for the extraction of lipids from yeast cells and its plasma membrane fraction. It can also be adopted to other membrane preparations of microbial origin.

1. Mix the harvested yeast cells (500 mg dry wt) or plasma membrane fraction (15–20 mg protein) with methanol (10 ml) and shake the suspension in a cell disintegrator (MSK-Braun Homogeniser) for two periods of 30 s at speed 2 (4000 vibrations per min) after addition of 35 g of glass beads (0.45–0.5 mm diameter).

2. Add chloroform to the suspension to give the ratio 2:1 of chloroform:methanol (v/v). Stir the suspension for 2 h on a flat bed stirrer at room temperature.

3. Filter the suspension using Whatman #1 filter paper. The extraction procedure can be repeated on the residue for quantitative extraction.

4. Transfer the extract to a stoppered glass tube and wash with 0.2 volumes of 0.9% NaCl to remove non-lipid contaminants.

5. Aspirate off the upper aqueous layer that has separated. Store the lower layer containing lipids at 4°C under nitrogen atmosphere for evaporation and concentration.

Extraction from other micro-organisms

1. Grow the organisms to the desired growth phase in a chemically defined medium.

2. Harvest the cells by centrifugation at 4000 rpm for 15 min and wash the cells three times with distilled water.

3. Resuspend the cells in propan-2-ol and heat at 70°C for 45 min, in order to inactivate degradative enzymes such as phospholipases (Hitchcock et al. 1986).

4. Let the suspension cool down to room temperature. Centrifuge the suspension to sediment the cells which are recovered after decanting the supernatant.

5. Resuspend the cells in 20 volumes of chloroform:methanol (2:1, v/v) and extract the lipids as follows:
   a) Transfer the cell suspension into a pre-weighed round-bottomed flask (RBF) and connect a condenser to the RBF.
   b) Reflux the suspension in the RBF for 30 min under N_2 gas (by connecting to the top of the condenser a balloon full of N_2 gas), keeping the temperature at 60–70 °C and avoid boiling.
   c) Filter the solvent mixture using a pre-weighed filter paper, washed previously with chloroform:methanol (2:1, v/v), and retain the filtrate (filtrate 1).
   d) Place the filter paper and the cells back into the pre-weighed RBF and extract with a fresh solvent mixture of chloroform:methanol [2:1, v/v following the same procedure as outlined in steps (a) and (b)].
   e) Filter the solvent mixture using another pre-weighed filter paper then add the filtrate (filtrate 2) to filtrate 1 obtained from step (c).
   f) Place the filter paper back into the RBF then dry the RBF, filter papers and cells overnight by placing in a desiccator at room temperature.
   g) Weigh the filter paper and cells and subtract from the weight of the RBF plus the filter papers. This should give the weight of cells.
   h) Evaporate the chloroform:methanol solvent mixture to dryness, obtained from the extractions, in a pre-weighed RBF using a rotary evaporator (temperature of the water bath should be around 70 °C).
   i) Place the RBF in a desiccator at room temperature for at least 30 min and weigh the RBF. The difference will give the weight of the crude lipids.

j) The weight of the dry cells will be the weight obtained in step (g) plus the weight obtained in step (i).

Evaporation and concentration of extracted lipids

This is a common step applicable to all lipid extracts obtained by following any of the above protocol. Basically, depending upon the volume of the extracted lipid, the evaporation can be done as follows:

Small volume. If the extracted lipid volume is small (10–15 ml), the solvent is best removed from it by evaporation under stream of N_2. One or two samples can be concentrated in a simple tube where N_2 can be bubbled slowly into the solvent while the tube is immersed in a water bath set between 65–70 °C.

In the case of too many samples, use thermostatically controlled heating blocks where N_2 can be passed into several samples through spaced needles simultaneously (Fig. 1).

Large volume. If the extracted lipid volume is large, the use of rotary evaporator (Fig. 2) for the evaporation of the solvent under reduced pressure and temperature is recommended. Such apparatuses are commercially available, and are connected to either water or vacuum pump. Using this device, the lipid extract can be continuously fed for evaporation and thus a large volume of it can be conveniently handled.

During evaporation, the lipid extract sometimes turns turbid due to the presence of residual water. In such instances, one can add 0.1 volume of benzene and continue with the evaporation. If necessary, this step can be repeated till a dry uniform film of lipid adhering to the walls of rotary flask, is obtained.

Purification of extracted lipids

Further purification of extracted lipids is necessary to analyse them by TLC/GLC/HPLC. The extracted lipids from any source can be purified as follows:

1. Dissolve the crude lipids in 100 ml of chloroform: methanol (2:1, v/v) and place in a separatory funnel.

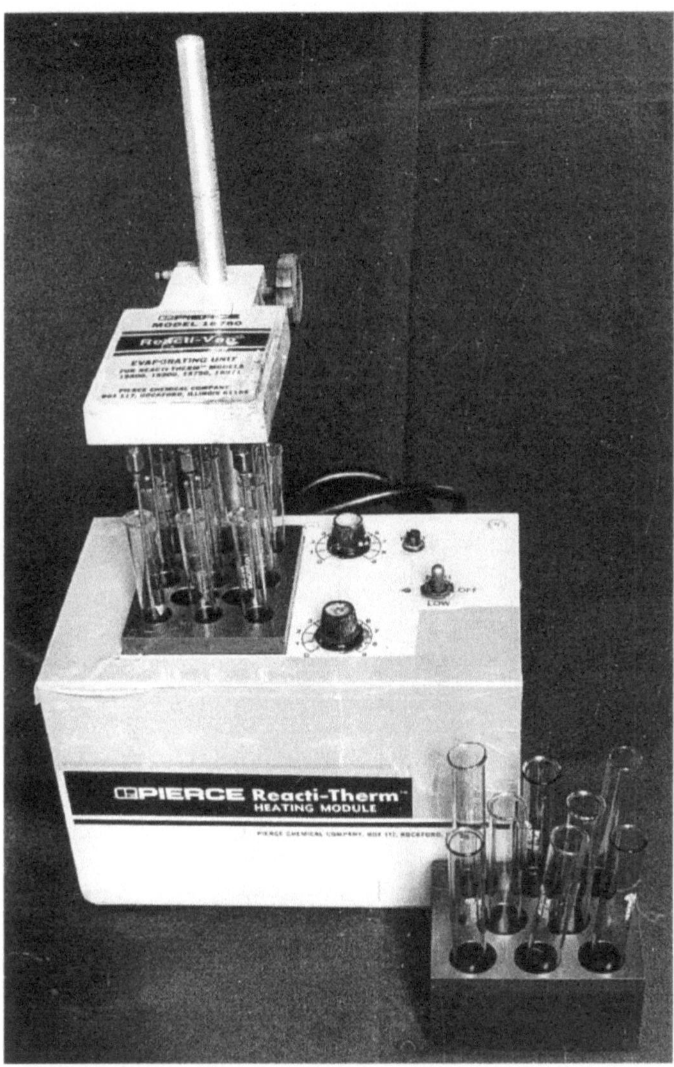

Fig. 1. Thermal block for multiple sample concentrator. Such commercially available blocks are good if small volumes of lipid extracts are to be concentrated. The solvent is evaporated under a stream of N_2, which is passed through the needles

Fig. 2. Rotary evaporator. It is good for concentrating large volumes of lipid extracts under reduced pressure and temperature

2. Add 30 ml of distilled water saturated with chloroform, to the separatory funnel. (To saturate the water with chloroform, add 5 ml chloroform to 30 ml water).

3. Shake gently in a horizontal manner and carefully release the pressure from time to time. Then let the separatory funnel stand for at least 10 min. The mixture will separate into two layers; the upper layer is an aqueous phase containing methanol and impurities, the lower layer is an organic phase containing chloroform and lipids.

4. Collect the lower phase (chloroform phase) in another separatory funnel and repeat steps 2 and 3 once more.

5. Combine the upper phases from the two extractions and re-extract with 30 ml chloroform using a separatory funnel.

6. Combine the two chloroform phases (step 4) and transfer into a pre-weighed RBF.

7. Evaporate the chloroform using a rotary evaporator (the temperature of the water bath should be around 70 °C).

8. Place the RBF at room temperature for at least 30 min in a desiccator containing anhydrous $CaCl_2$. Then weigh RBF. The difference will give the weight of the pure lipids.

9. Dissolve the pure lipids in chloroform (10 mg of pure lipids should be dissolved in 1 ml chloroform) and transfer into a glass vial.

10. Store the glass vial under N_2 gas (to prevent possible oxidation of the lipids) at −20 °C until needed.

3   Removal of Non-Lipid Contaminants

Lipid extracts have a tendency to trap non-lipid material because phospholipids tend to form micelles. The non-lipid contaminants are water-soluble substances such as sugars, amino groups, urea and salts. Several methods are known for the removal of non-lipid contaminants, e.g. removal of non-lipids by solvent partitioning, dialysis, evaporation and re-extraction, Sephadex chromatography, coprecipitation with protein etc.

Most commonly, the non-lipid contaminants present in lipid extract can be removed by simply washing the extract with 0.2 volume of water or 0.9% NaCl solution. Calcium and magnesium salts should not be used, as they result in low recoveries of acidic phospholipids. The technique suffers from the disadvantage that it is difficult to apply to large volumes of extract, as there is some loss of lipid, particularly that of gangliosides, into the upper layer (gangliosides, if needed, can be separated from the upper phase by dialysis).

The more elegant and complete, though more time-consuming, method of removing non-lipid contaminations is to carry out the washing procedure by liquid/liquid partition chromatography on a column. The method essentially consists of passing the lipid extract through a liquid/liquid partition column. Sephadex G-25 matrix is used to immobilise the aqueous phase which contains non-lipid contaminants and gangliosides.

Removal of non-lipid contaminants by dialysis is not recommended because there is no simple way of knowing when it is complete.

4   Storage of Lipids

Standard solution of extracted lipids

The extracted dry lipid can be resuspended in known volume of chloroform or, if necessary, lipid can be weighed in a sensitive balance and a known weight by volume solution (e.g. 10 mg/ml) in chloroform can be made. The dissolved lipid is stored in cold under nitrogen atmosphere, which can be maintained by flushing the nitrogen into the container followed by tight sealing. This solution is then ready for lipid analysis and quantitative estimations of different class of lipids.

Storage

Lipids should be stored in glass containers. General contamination can be reduced by not greasing glass tops, by using Teflon taps for separatory funnels and columns and by sealing desiccators with silicone rubber. Corks and rubber stoppers should not be used in lipid work, even polyethylene stoppers can cause contamination and allow solvent to escape.

Since most common lipids contain unsaturated fatty acids, care must be taken to avoid auto-oxidation of the sample during storage. Auto-oxidation can be minimised by working with oxygen-free solvents and by performing all manipulations under nitrogen atmosphere.

Purified lipid extracts may be stored in tightly closed vials or in a container wrapped in aluminium foil at low temperature ($-20\,°C$ or lower) in the presence of inert solvents and gases, for short periods.

If lipid solution has to be stored for a longer period, <0.005% of antioxidants such as 2,6-di-tert-butyl-p-cresol (or butyl hydroxy toluene, BHT), are added to it, which effectively prevents oxidative degradation of unsaturated lipids. The antioxidant can be removed by chromatography.

5   Tips, Tricks and Precautions

- During the extraction procedure, contamination from solvents may occur. In order to avoid this, very pure solvents (preferably redistilled) should be used.
- Halide-containing solvents and alcohols have been shown to be quite sensitive to light. Hence solvents, in general, should be stored in brown bottles.
- Plastic bottles or tips should not be used during extraction procedure. Although they seem to be inert, they contain oxidants and low molecular weight polymers, which can enter the solvents. Preferably, glass apparatus for the extraction of lipids and glass pipettes for measuring lipid extracts should be used.
- If highly unsaturated lipids are to be extracted, it is recommended to minimise lipid peroxidation by removing dissolved oxygen by slowly flushing nitrogen gas through the solvent.
- The ideal solvent or solvent mixture for extracting lipids from tissues should be sufficiently polar to remove all lipids, but it should not be so polar that triacylglycerols and other

non-polar simple lipids do not dissolve and thus are poorly extracted.

- Care should be taken if diethyl ether is used because of its ease of auto-oxidation and the possibility of the presence of ether peroxides in the original solvent. Some brands of anhydrous diethyl ether contain antioxidants, and normally this is indicated on the label; however, BHT found in certain brands is not disclosed.
- Polar lipids are sparingly soluble in hydrocarbon solvents, but they dissolve readily in more polar solvents such as methanol, ethanol or chloroform. It is seen that the shorter the chain length of the fatty acid residue, the greater is the solubility of the lipid in more polar solvent.
- Diethyl ether and chloroform are good solvents for lipids, but with this solvent mixture, complex lipids from tissues cannot be extracted. Propan-2-ol: hexane mixture has been recommended to remove lipids from animal tissue.
- Many solvents are toxic and even carcinogenic. Care should be taken to avoid body contact. If mixture of chloroform and methanol spills over a part of body, it causes a severe burning sensation. The exposed part should immediately be kept under cold running water to minimise the burning sensation, which could last for about 10 min.
- During extraction of lipids from dry samples, water may be added to help in extraction of lipids. Most enzymes are denatured by mixtures containing methanol, but phospholipase activity in plants is often enhanced by the presence of ether or methanol and loss of complex lipids may result. The enzyme may be denatured by blanching the tissues in boiling water or hot methanol or by hot propan-2-ol extraction before treatment with chloroform and methanol.

Table 1 lists common solvents which have been arranged in order of increasing dielectric constant (increasing polarity).

Table 1. Common solvents arranged in order of increasing polarity

Solvent	Dielectric constant
Pentane	1.80
n-Hexane	1.89
Heptane	1.92
Cyclohexane	2.02
Carbontetrachloride	2.02
Benzene	2.28
Diethylether	4.34
Chloroform	4.80
Ethylacetate	6.02
Pyridine	12.30
Acetone	20.70
Ethanol	24.30
Methanol	33.60
Water	80.37

References

Folch JLM, Stanley GHS (1957) A simple method for the isolation and purification of total lipids from animal tissues. J Biol Chem 226:497–509

Hitchcock CA, Barrett-Bee KJ, Russell NJ (1986) The lipid composition of azole-resistant strains of *Candida albicans*. J Gen Microbiol 132:2421–2431

Nichols BW (1963) Separation of the lipids of photosynthetic tissues: improvement in analysis by thin layer chromatography. Biochim Biophys Acta 70:417–422

Chapter IV   Chromatographic Analysis of Lipids

Ashraf S. Ibrahim[1] and Mahmoud A. Ghannoum[1,2]

1   Background

Generally the same methods are applicable in chromatographic analyses of lipids, irrespective of their origin. Therefore, we describe the chromatographic techniques routinely used in our laboratory for the lipid analyses of microbial origin. Although departure from the methods described herein may be necessary at times, the techniques described in this chapter should, however, be of general use for thin layer and gas liquid chromatographic analyses of membrane lipids.

2   Analysis of Lipids Using Thin Layer Chromatography (TLC)

The extracted lipids contain polar as well as apolar lipids (see Chaps. I and III, this Vol.). Different methods are employed for the separation and analysis of different classes of lipids. These include solvent fractionation, column chromatography, thin layer chromatography (TLC), gas liquid chromatography (GLC) and high performance liquid chromatography (HPLC).

[1] Division of Infectious Diseases, Department of Medicine, Harbor-UCLA Medical Center, St. John's Cardiovascular Research Center, Bldg R-B2, 1000 W. Carson St., Torrance, California 90509, USA
[2] The UCLA School of Medicine, Los Angeles, California 90024, USA

This section discusses separation of lipids by TLC. This technique is simple, versatile and highly sensitive with the flexibility to be used both qualitatively and quantitatively.

2.1 Separation of Lipids

TLC is carried out by spreading a layer of an adsorbent on a support such as glass, plastic or aluminium sheet. The adsorbent could be silica gel, alumina, cellulose or sepharose. Lipid analysis is usually carried out using silica gel. There are two types of silica gels: silica gels which contain $CaSO_4$ as binders (type G) and those without $CaSO_4$ binders (Type H). A variety of prepared plates can also be obtained from commercial sources. Unless unusual adsorbents are required or many plates are needed, it is convenient to buy pre-prepared plates. TLC glass plates are frequently used, since they can be cleaned and used again. In order to clean the glass plates, the adsorbent layer is removed by washing with tap water, and immersed in 10% solution of alkali for several hours. The plates are then washed with distilled water and dried in a drying oven. Unless otherwise stated, the separations should be carried out using 20×20 cm glass plates coated with silica gel G (thickness of 0.25 mm). It is advisable to activate the prepared TLC plates by heating at 110 °C for 30–60 min. After activation the plates are cooled in a desiccating cabinet before applying the sample. The activation step is very important, particularly under humid weather conditions.

Thin layer plates

Rectangular glass tanks are used to run the TLC plates. These tanks are saturated with the solvent system being used. This is done by pouring sufficient amount of the solvent mixture (to a depth of 1.5 cm) followed by immersing filter paper cuts in the solvent system to fit three sides of the tank, leaving one of the larger sides without filter paper to monitor the chromatoplate during the separation process. The lid of the tank is sealed with

Tanks

petroleum jelly and removed for the minimum time necessary to place the plate in the tank in order to keep the atmosphere of the tank saturated with the solvent mixture used. Saturated tanks prevent evaporation of the solvent from the adsorbent layer during ascending chromatography, thus ensuring good separation of the lipids. The saturation process should be done at least 1 h before developing the chromatoplates.

Sample application in unidimensional TLC

Lipid samples are dissolved in a polar solvent, such as chloroform or chloroform:methanol (1:1, v/v). The samples are applied using either a microsyringe with a volume between 10–50 µl (available commercially) or micropipette. In analytical TLC, samples are applied as spots, 3–5 mm in diameter, 2–3 cm from the bottom edge of the plate. If the lipid mixture is dissolved as 10 mg/ml in chloroform, then applying 50 µl (500 µg lipid mixture) is sufficient to run the TLC. A fine-spotting line is drawn with a pencil (making sure not to damage the adsorbent layer). This line will help the worker in positioning the spots at the same level. There are also commercially available templates to guide the sample spotting. The sample is applied as one spot after the other on the same location, allowing the solvent to evaporate before the next application. Care should be taken not to damage the adsorbent layer, since this causes irregularity in the movement of the solvent, which may result in a poor separation. It is advisable to run standards on the same plate since the R_f values show some variation between plates run on different days. The standards should be pure in order to avoid misinterpretation of the chromatograms. About 10 µg of the pure standard in 10 µl of the solvent should be applied in a single spot next to the sample (Fig. 1).

Sample application in Two-dimensional TLC

In certain lipid classes (e.g. phospholipids) using a unidimensional system does not completely resolve the lipid mixture. In such cases, it is necessary to use two-dimensional solvent systems run at right angles to each other to separate the lipid mixture. Only one sample is run on a single plate. Identifica-

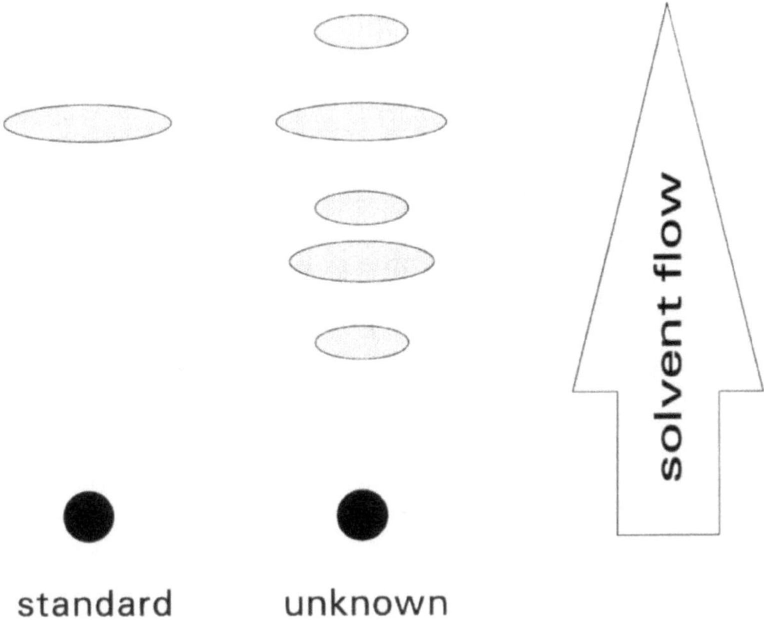

Fig. 1. Application of the samples in unidimensional TLC. The *arrow* indicates the direction of solvent movement

tion of different lipid fractions is done by comparing the separated spots with standards run using the same solvent system. Usually 2 mg of the sample is applied as a spot around 2–3 cm from the right bottom edge of the plate and the plate is developed in such a direction so that the spots separated are on one edge of the plate. The plate is dried and turned 90° clockwise (so that the lipid spots are along the bottom). The plate is then developed in another solvent system (Fig. 2).

After applying the sample and the standards, the TLC plate is left at room temperature for 1–2 min to allow the residual chloroform to evaporate. Then the chromatoplate is placed in the tank containing the solvent system required and left there till solvent front reaches about 1 cm from the top of the plate.

Development of chromatograms

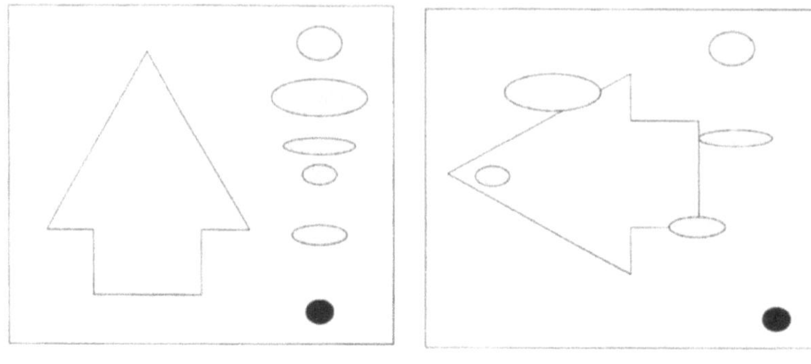

Fig. 2. Two-dimensional TLC. The lipids to be separated are spotted on the *lower right hand corner* of the plate. After developing in the first direction, the plate is rotated 90° clockwise then developed in the other direction. The *arrows* indicate the direction of solvent movement

! **Note.** Only the purest solvents are used in the development of the plate, and even those should be redistiled freshly.

Separation of fungal lipids using TLC The chromatographic analyses of fungal lipids (as an example of the procedures used in chromatographic analysis of lipids), are described below. The lipid content of many fungal species has been studied in detail (Weete 1980). Fungal lipids vary tremendously, depending on the culture conditions and the species studied. In general, lipids of fungi are divided into three categories: polar lipids (e.g. phospholipids and glycolipids), apolar lipids (e.g. triacylglycerols free fatty acids and sterols) (see Chaps. I and III, this Vol.). No single solvent system will separate all lipid classes. Therefore, it is important to use different solvent systems to comprehensively analyse the lipid classes or to choose a particular solvent system, depending on the aim of the study.

Selection of the solvent system for lipid separation depends on the overall polarity of the solvent mixture and its acidic/ basic nature. Several solvent systems are mentioned below.

These solvent systems are selective for the major lipid classes and are not exclusive. Other solvent systems are listed in Table 1.

Separation of apolar lipids. Apolar lipids are fractionated on TLC plates coated with silica gel G, using the solvent system of hexane:diethylether:acetic acid (90:10:1 or 75:25:1, v/v). This solvent system separates the apolar lipids into monoacylglycerols, sterols, diacylglycerols, fatty acids, triacylglycerols, alkyl esters, and steryl esters (Fig.3). In this system polar lipids remain at the origin.

Separation of polar lipids. Polar lipids are resolved unidimensionally on silica gel G using the solvent system of chloroform:methanol:7N ammonium hydroxide (65:25:4, v/v). This solvent system results in a limited separation of the different phospholipid classes (Fig. 4). It is necessary to use a two-dimensional TLC for better separation. Chloroform:methanol:7N ammonium hydroxide (65:25:4, v/v) is used as the solvent system in the first direction followed by chloroform:methanol:acetic acid:water (170:25:25:4, v/v) in the second direction (Fig. 5).

2.2   Detection

After running the TLC plate in the desired solvent system, the plate is allowed to dry in a fume cupboard for at least 30 min, then the separated lipid classes are visualised by spraying the chromatoplate with different reagents. This is carried out by using a sprayer bottle powered by an aerosol that can be purchased from commercial sources. Two types of detection tests should be performed to detect lipid spots on a TLC plate:

- General tests, which reveal all lipid spots.
- Specific tests, each of which reveals a group of lipid classes or even an individual class of lipids.

Table 1. Some of the solvent systems used in the separation of membrane lipids

Neutral lipid separation
  Hexane:diethylether:glacial acetic acid (90:10:1, v/v)
  Hexane:diethylether:glacial acetic acid (75:25:1, v/v)
  Heptane:diethylether:glacial acetic acid (60:40:1, v/v)

Unidimensional separation of phospholipids
  Chloroform:methanol:7 M ammonia (65:25:4, v/v)
  Chloroform:methanol:glacial acetic acid:water (25:15:4:2, v/v)
  Chloroform:methanol:glacial acetic acid:water (60:50:1:4, v/v)
  n-propanol:propionic acid:chloroform:water (3:2:2:1, v/v)

Two-dimensional systems for phospholipid separation

First dimension	Second dimension
Chloroform:methanol:7 M ammonia (65:30:4, v/v)	Chloroform:methanol:acetic acid:water (170:25:25:4, v/v)
Chloroform:methanol:ammonia (65:35:5, v/v)	Chloroform:methanol:acetone: acetic acid:water (10:2:4:2:1, v/v)
Chloroform:methanol:7 M ammonia (90:54:11, v/v)	Chloroform:methanol:acetic acid:water (90:40:12:2, v/v)
Chloroform:methanol:acetic acid:water (50:20:7:3, v/v)	Chloroform:methanol: 40% aq. methylamine:water (13:7:1:1, v/v)

Acidic phospholipid separation
  Solvent system I: acetone:light petroleum (1:3, v/v)
  Solvent system II: chloroform:methanol:glacial acetic acid:water (80:13:8:0.3, v/v)
  Both solvent systems are used in the same dimension

Polyphosphoinositide separation
  n-propanol:4 M ammonia (2:1, v/v)
  Chloroform:methanol:4 M ammonia (9:7:2, v/v)
  Chloroform:methanol:28% ammonia:water (40:48:5:10, v/v)

Ganglioside separation
  Chloroform:methanol:2.5 M ammonia (60:40:9, v/v)
  Propanol:water (7:3, v/v)

Several methods are used to detect lipids without discriminating between different lipid classes. The most commonly used methods are iodine vapour and sulphuric acid charring.

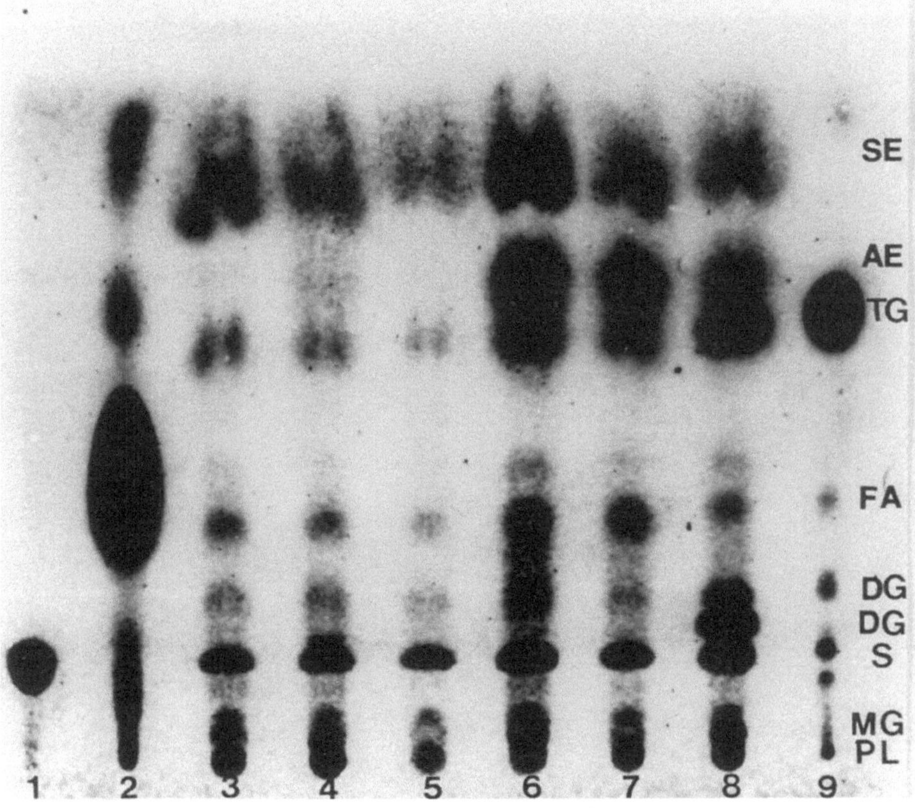

Fig. 3. Typical TLC plate showing apolar lipids extracted from *Candida albicans*. The adsorbent was silica gel G; the solvent was hexane: diethylether:acetic acid (90:10:1, v/v). The lipids were visualised by charring. *1* Sterol standard; *2* fatty acid standard; *3–8* different lipid samples extracted from *C. albicans*; *9* triacylglycerol standard; *SE* steryl esters; *AE* alkyl esters; *TG* triacylglycerols; *FA* fatty acids; *DG* diacylglycerols; *S* sterols; *MG* monoacylglycerols; *PL* polar lipids

Detecting lipid spots using iodine vapour is the most rapid and non-destructive method. After letting the chromatoplate dry from the developing solvent in a fume cupboard, the chromatoplate is placed in another covered developing tank containing crystals of iodine. After a few minutes to several **Detection using iodine vapour**

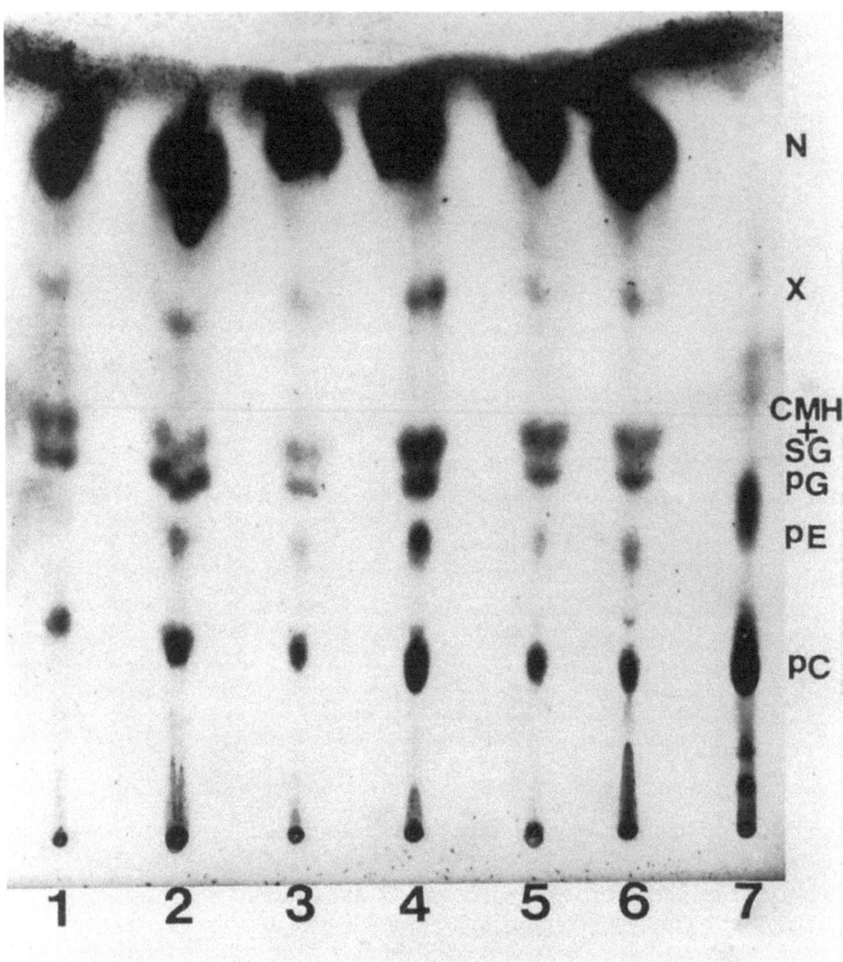

Fig. 4. Typical TLC plate showing polar lipids extracted from *Candida albicans*. The adsorbent was silica gel G; the solvent was chloroform: methanol:7 M ammonia (65:25:4, v/v). The lipids were visualised by charring. *1–6* different lipid samples extracted from *C. albicans*; *7* phospholipid standard; *N* neutral lipids; *X* glycolipids; *CMH* ceramide monohexosides; *SG* steryl glycosides; *PG* phosphatidylglycerols; *PE* phosphatidylethanolamines; *PC* phosphatidylcholines

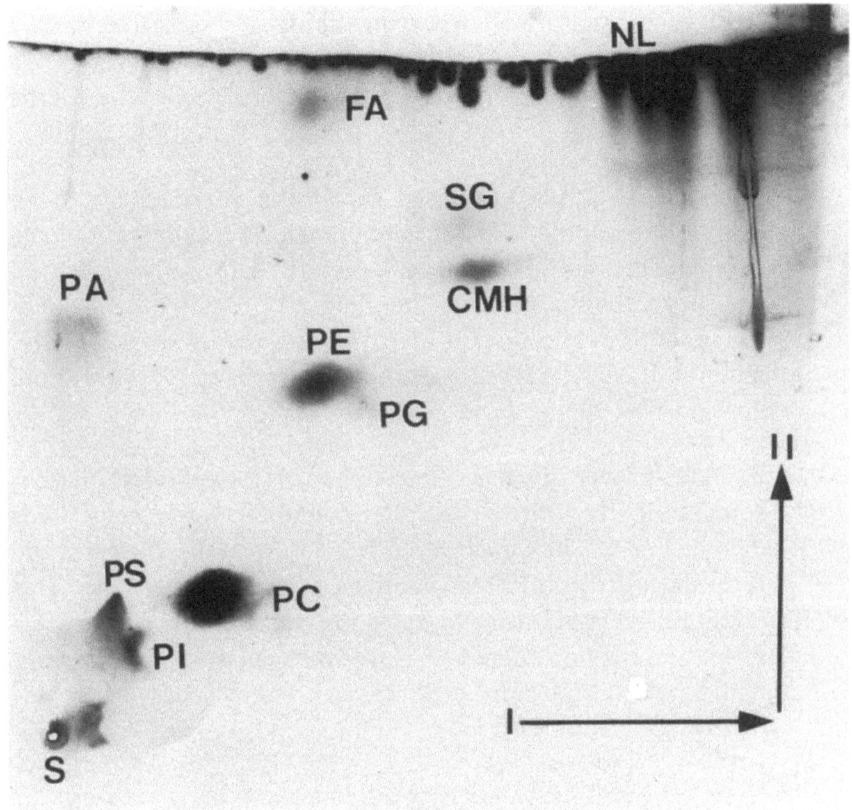

Fig. 5. Typical two-dimensional TLC plates of polar lipids isolated from *C. albicans*. The adsorbent was silica gel G. The solvents were: *I* chloroform:methanol:7M ammonia (65:30:4, v/v) and *II* chloroform: methanol:acetic acid:water (170:25:25:4, v/v). The lipids were visualised by charring *PI* Phosphatidylinositols; *PS* phosphatidylserines; *PA* phosphatidic acids; *PC* phosphatidylcholines; *PE* phosphatidylethanolamines; *PG* phosphatidylglycerols; *CMH* ceramide monohexosides; *SG* steryl glycosides; *FA* fatty acids; *NL* neutral lipids

hours (depending on the amount of the separated lipids), brown spots appear on a yellow background.

For further analysis of the individual spots, the lipid spots should be outlined using a pencil before the spots start fading.

Most of the naturally occurring lipids are sensitive to this test. However, certain lipids (e.g. completely saturated lipids and glycolipids of animal origin) are not detected by iodine vapour.

! **Note.** If the lipids visualised by iodine vapours are to be counted for radioactivity, it is important to let the iodine fade completely, since it quenches, before using a scintillation counter. Usually, warming the plate accelerates the fading of the spots. If the lipid spots are to be assayed by a colorimetric method, then the presence of residual traces of iodine will not pose a problem.

Detection using sulphuric acid charring Using a sprayer under a fume cupboard, the chromatoplate is sprayed lightly with 50% sulphuric acid (v/v) and the plate is then heated at 220 °C in an oven for 15 min. All lipid classes will form dark brown or black spots on a white background. This method is very sensitive in detecting lipids and is easy to perform. The only problem is that this method does not discriminate between lipids and other organic compounds which may be present on the TLC plate.

! **Note.** Care should be taken when removing the plate from the hot oven and placing it on a cold surface. This act may lead to glass breakage.

Detection of phospholipids The phosphate-containing lipids can be detected using the method described by Dittmer and Lester (1964). The stain is prepared as follows:

a) 40.1 g of MoO_3 is dissolved in 1 liter of 25 N H_2SO_4 by boiling gently.
b) 1.78 g of powdered molybdenum is added to 500 ml of reagent (a), and the mixture is boiled gently for 15 min. The solution is cooled and decanted.
c) Equal volumes of reagent (a) and reagent (b) are mixed and the combined solution is mixed with 2 volumes of water.

This final reagent (molybdenum blue reagent) has a greenish yellow colour and is stable for months.

The chromatoplate is sprayed evenly with reagent (c). After a few minutes, the phospholipid will appear as dark blue spots on a white, or light blue-grey background. The neutral lipids and glycolipids do not give positive tests.

This test specifically stains phosphatidylcholine and choline-containing sphingomyelin. The reagent is prepared as follows:

a) 40 g potassium iodide is dissolved in 100 ml distilled water
b) 1.7 g bismuth subnitrate is dissolved in 100 ml of 20% acetic acid
c) 5 ml of solution (a) is mixed with 20 ml of solution (b), then diluted to 75 ml to give the desired spray reagent

Detection of choline-containing phospholipids (Dragendorff's test)

The TLC plate is sprayed with reagent (c). The choline-containing lipids will appear as orange-red spots

Lipids containing free amino groups (e.g. phosphatidylethanolamine, phosphatidylserine) are detected using ninhydrin spray. The TLC is sprayed with 0.25% ninhydrin dissolved in acetone. After heating the plate at 100 °C for 5–15 min the amino-containing lipids appear as pink-purple spots.

Detection of amino group containing lipids (ninhydrin test) !

Note. It is advisable to wear gloves when using the ninhydrin spray.

Detection of glycolipids

α-Naphthol specifically stains glycolipids. The stain is prepared as follows:

a) 0.5 g of α-naphthol is dissolved in 100 ml of methanol:water (1:1, v/v)
b) 50% H_2SO_4

The TLC plate is sprayed with α-naphthol reagent (a) until the plate is wet and it is then left to dry. Next, the plate is sprayed with 50% sulphuric acid (b) and heated in an oven at 220 °C for 20 min. Glycolipids appear as purple spots.

2.3 Quantitative Estimation of Phospholipids

In order to quantify phospholipids, its phosphorous is estimated after converting it to inorganic phosphate (P_i), which is achieved by hydrolysing the phospholipids by an anhydrous acid. Following the release of phosphorous, it is estimated by any method of P_i estimation. Below, a typical protocol to estimate lipid phosphorous is described (Wagner et al. 1962):

Reagents
a) Perchloric acid (60%)
b) Ascorbic acid (10%)
c) Ammonium molybdate (2.5%)
d) Perchloric acid (5%)
e) Standard phosphate solution: 3.58 g of $Na_2HPO_4.12H_2O$ dissolved in 1 l H_2O (10 mM). Store frozen. Dilute 1:10 before use (1 µmol/ml).

Colouring reagent
Prepare by mixing b:c:d (1:1:8, v/v) just before use.

! **Note.** Preferably ascorbic acid (b) should be added after thorough mixing of (c) and (d).

Procedure
1. Scrap the phospholipid spot from the TLC plate using a sharp razor blade and transfer carefully to good-quality tubes (Pyrex). This is done immediately after visualising the phospholipid by iodine vapours. It is not necessary to remove the iodine stain for this purpose.

! **Note.** Scrap silica gel rom an area of plate where there is no detectable spot. This tube is used as a blank of silica gel.

2. Add 0.5 ml of perchloric acid (60%) to each tube. Digest the contents of tubes at 180 °C in a heating block or sand bath for 2 h.

Note. This should be done under a fuming cupboard as acid **!**
digestion will give very irritating fumes. After the completion of digestion, the contents of each tube should turn white or light yellow. If some tubes still have dark contents, they may be heated for an extra time.

3. Cool the contents of the tube and add 5 ml of freshly prepared colouring reagent to each tube. Incubate the tubes at 37 °C for 2 h in a water bath shaker.

4. The blue colour developed can be read at 660 nm against a blank prepared under identical conditions.

Note. It is advisable to centrifuge the contents of tube to settle **!**
silica gel.

5. Prepare a standard curve using different concentrations of solution (e). It is not necessary to digest and centrifuge these samples.

6. The μmoles of phospholipids can be determined directly by reading the absorbance from the standard curve.

Note. The phosphorous of phospholipids is measured directly **!**
by taking an aliquot of the lipid extract and evaporating it to dryness in the tubes which could then be directly used for digestion. Add 0.5 ml of (a) and follow steps 1–6, but omit the centrifugation step, since there would be no silica gel in these samples.

3   Analysis of Fatty Acids Using Gas Liquid Chromatography (GLC)

3.1   Derivatisation of Fatty Acids

Fatty acids of total lipids extracted from fungi can be analysed using GLC after derivatisation. This is carried out by a process called methanolysis, which converts the fatty acids into the methyl ester form (volatile form). The fatty acids are derivatised then purified by the following procedure:

Reagents
- 1% Sulphuric acid
- Diethylether
- Chloroform

Derivatisation
1. For each 2 mg total lipids add 4 ml of 1% sulphuric acid in absolute methanol.

2. Heat the mixture under nitrogen atmosphere at 90 °C for 90 min (prevent drying by adding absolute methanol).

3. Let the reaction mixture cool down to room temperature, then add 5 ml of distilled water.

4. Extract the methyl esters by adding 3 ml of diethylether and shake the tube gently. Let the tube stand till the mixture separates into two phases.

5. Collect the upper layer (organic layer containing methyl esters dissolved in ether) with a Pasteur pipette. This step has to be done carefully without withdrawing any of the aqueous phase.

6. Repeat five times steps 4 and 5, combining the upper layer from each extraction.

7. Evaporate the diethylether using N_2 gas under a fume cupboard, then redissolve the methyl esters in 200 μl of chloroform.

The dissolved methyl esters are further purified using TLC **Purification** prior to GLC analysis: **of fatty acid methyl esters**

1. Using a micropipette, spot about 10 µl of the methyl ester sample on the left side of the TLC plate (size 20 × 20 cm).

2. Apply the rest of the sample parallel to the 10 µl sample (Fig. 6).

3. Separate the fatty acid methyl esters by running the TLC plate in hexane:diethyl ether (80:20, v/v).

4. When the solvent front reaches about 1 cm form the top, take the chromatoplate out and allow it to dry.

5. Cover the plate with a piece of the paper in such a way as to cover the whole plate except the part where the 10 µl sample was spotted.

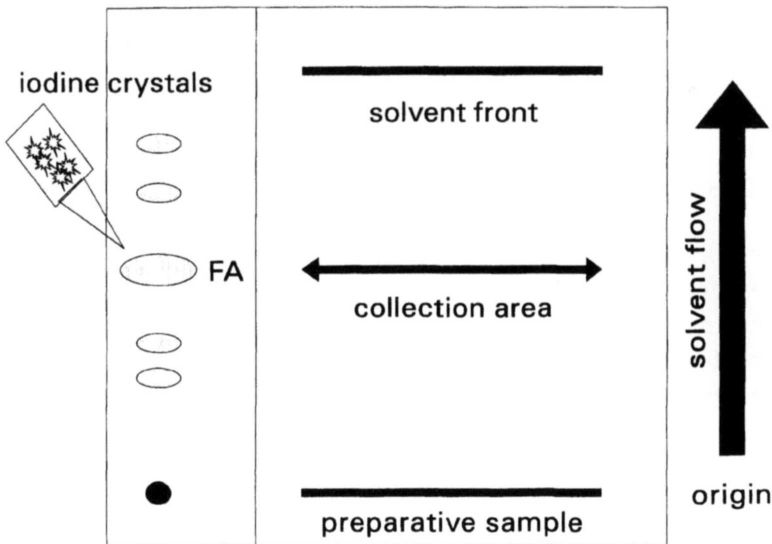

Fig. 6. Preparative thin layer chromatography showing the purification method of fatty acid methyl esters. *FA* Fatty acid methyl esters

6. With the help of a 10-ml pipette containing a few crystals of iodine (Fig. 6), start blowing air in the pipette, directing its end towards the 10-µl spot and all the way up to the upper edge. The fatty acid methyl esters will appear as a spot almost in the middle of the TLC plate.

7. Immediately, before the colour fades away, outline the spot showing the methylated fatty acid with a pencil.

8. Scrap the silica gel which is parallel to the visualised spot and perpendicular to the spot of the rest of the sample (Fig. 6). Scraping is carried out by using a scraper or a razor blade.

9. Transfer the scraped silica gel to a glass tube and extract with 3 ml diethyl ether.

10. Repeat step 9 four times, combining the ether layer each time, then evaporate the ether using N_2 gas under the fume hood.

11. Store the purified methyl ester sample at $-20\,°C$ till analysed by the GLC.

3.2 Derivatisation of Fatty Acids of Individual Lipid Classes by Methanolysis

Reagents
- Chloroform
- Methanol
- Ammonium hydroxide
- Hexane
- Diethyl ether
- Acetic acid

Procedure
1. Spot 50 µl of the fatty acid sample, at about 2 cm of the left edge of a TLC plate.

2. Parallel to the spot made in step (1), spot 10 mg of the total lipid across the plate (1 ml).

3. Put the TLC plate in a developing tank containing one of the following solvent systems:

   - Chloroform:methanol:ammonium hydroxide (65: 30:4, v/v) to separate the polar lipids.

   - Hexane:diethylether:acetic acid (75:25:1, v/v) to separate the apolar lipids.

4. After running the TLC plate, allow it to dry, then cover the plate with a piece of paper and blow iodine vapour on the 50 µl spot as explained earlier (step 5 and 6 of Sect. 3.1, Purification of Fatty Acid Methyl Esters).

5. Identify the spots and outline them immediately with a pencil.

6. Scrap the silica gel areas parallel to each spot and transfer into glass tubes.

7. Derivatise the lipid classes adsorbed on the silica gel by following the steps mentioned in Section 3.1, Derivatisation.

These purified fatty acid methyl esters are used for analysis by GLC.

3.3 Analysis of Fatty Acids by GLC

1. To analyse the purified fatty acid methyl esters using GLC, add 50 µl of hexane to the vial containing the fatty acid methyl esters. **Gas liquid chromatography**

2. Using a gas chromatography syringe inject 1 µl of the hexane dissolved mixture into a Diethylene Glycol Succinate (DEGS, 15% on 80/100 chromosorb W-HP) column.

3. The GLC conditions should be adjusted to the following conditions:

   - Oven or column temperature = 180 °C

- Injector temperature $= 250\,^{\circ}\mathrm{C}$

- Flame ionisation detector (FID) temperature $= 300\,^{\circ}\mathrm{C}$

4. Individual fatty acids are identified by comparing the retention time of the separated fatty acids with retention times of commercially available fatty acid standards. Figure 7 is an example of a chromatogram of fatty acids analysed by GLC.

3.4   Description of the GLC System

GLC is an extremely versatile technique which is widely used in lipid analysis. The machine is very sensitive with high resolution, provides fast analysis (5 to 30 min for most samples) and is easy to operate.

! **Note.** There are certain limitations to the use of GLC. These are: the samples to be analysed must be volatile, dirty samples require cleaning up, use of other instruments for confirmation of peak identity may be necessary, some training or experience in running the GLC instrument is required and some classes of compounds cannot be analysed using this technique (e.g. high molecular weight polymers, wood and plastic).

Similar to other chromatographic techniques, GLC samples are separated by partition between a mobile and a stationary phase. The mobile phase is an inert gas (called the carrier gas since it carries the injected sample through a column) e.g. nitrogen or helium, and the stationary phase is a solid stationary support coated with a liquid phase of DEGS which is the most commonly used liquid phase for fatty acid separation. The stationary phase is packed into a glass or a stainless steel column. One end of the column is attached to an injector and the other end to a detector (Fig. 8).

There are several detectors used for separating different samples. Flame ionisation detector (FID) is used for lipid and sterol analysis. It is beyond the scope of this chapter to discuss

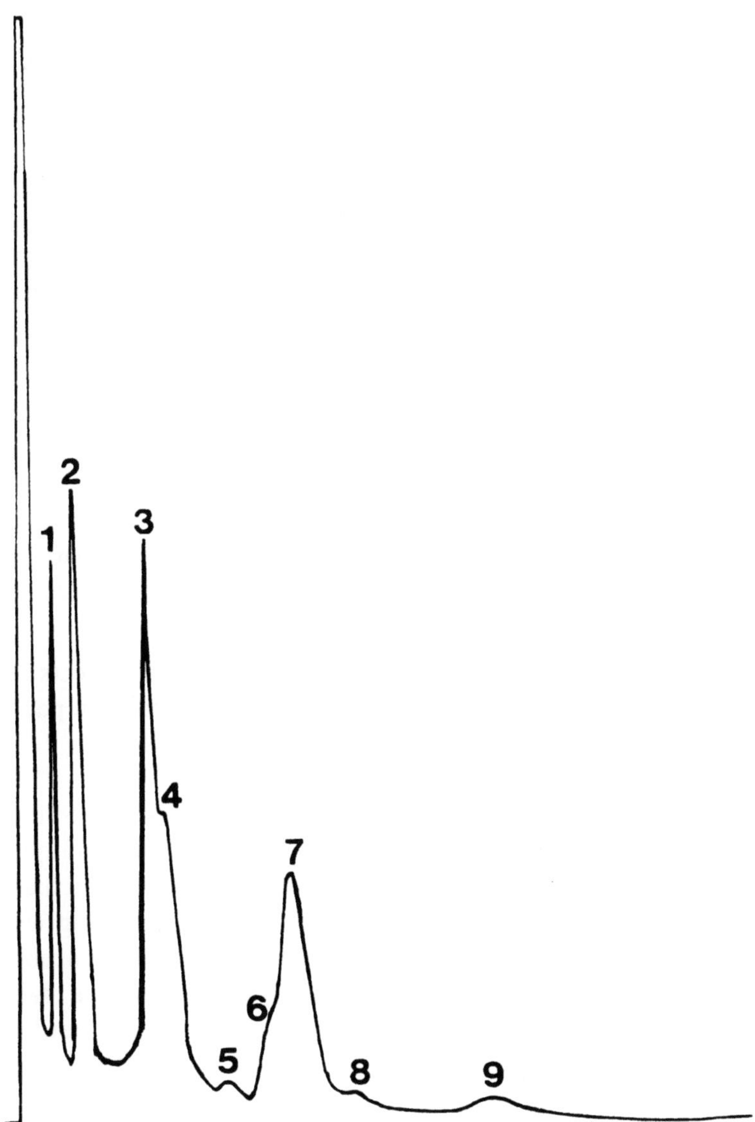

Fig. 7. Gas liquid chromatograms on 15% DEGS of methyl esters of fatty acids isolated from a bacterial culture. *1* Lauric ($C_{12:0}$); *2* myristic ($C_{14:0}$); *3* palmitic ($C_{16:0}$); *4* palmitoleic ($C_{16:1}$); *5* stearic ($C_{18:0}$); *6* tuberculostearic; *7* oleic ($C_{18:1}$); *8* linoleic ($C_{18:2}$); *9* linolenic ($C_{18:3}$) acids

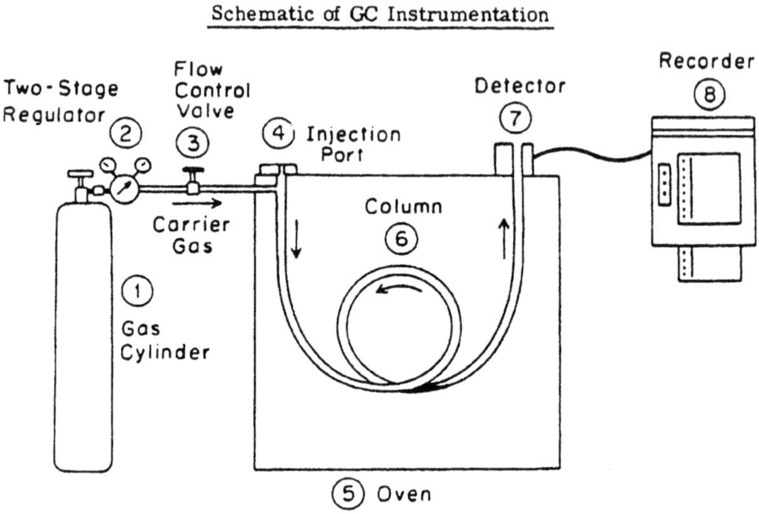

Fig. 8. Scheme of gas liquid chromatography (GLC) instrumentation

GLC in detail; however, we describe briefly several aspects related to the preparation of the stationary phase, column packing and syringe handling.

Preparation of column supports In gas liquid chromatography, the solid support is packed into a column. This solid stationary support is coated with a liquid phase. The solid supports are inert porous material which absorb large quantities of liquid phase without appearing wet. The most common supports are made from diatomaceous earth. They can be purchased in a variety of mesh ranges ready for use (some workers prefer additional treating and/or sieving). The most common mesh range is 60/80 (the higher the number, the finer the particle.

The liquid phase must have a very low pressure at the operating temperature and should possess some differential selectivity for the components of the mixture to be separated. It is usually applied up to 40% by weight and the most common range is up to 20% of liquid phase.

To ensure a uniform coating by the liquid phase on the solid support, the former is usually dissolved in a volatile solvent, then slurred with the solid, finally heated and stirred gently to remove the solvent.

GLC columns are usually made of metal or glass, 1/4 or 1/8 inch in outer diameter (o.d.) and 2 to 6 ft long. Packing short columns (2 to 6 feet) is easy, and these lengths are quite common. A small wad of Pyrex glass wool is used to close one end of the column. **Packing a column**

The prepared column packing is added in small amounts with some form of agitation. The column can be "vibrated" or "tapped" (e.g. with a spatula). When the column is full, another glass plug is used to close the other end, and fittings are attached.

The column is "preconditioned", prior to separation of the samples. Preconditioning a new column is accomplished by temperature programming the GLC to heat the column starting at a slow rate with a steady increase in the temperature to reach the maximum column temperature, then keeping the column at this temperature for a period of time, depending on the liquid phase used.

For liquid samples, the most common sampling device is a syringe. The size of the most commonly used syringes range from 1 to 50 µl. The syringe is filled with the sample which is injected through a self-sealing septum into a heated zone for quick volatilisation (alternatively, the sample can be injected directly onto the column). Common sample volumes injected are 1 to 10 µl. **Syringe techniques**

When filling the syringe, exclude all the air prior to drawing the sample. This is accomplished by repeatedly drawing an aliquot of the sample into the syringe and rapidly expelling it. Then slowly draw up about twice as much sample into the syringe as you plan to use. Hold the syringe vertically with the needle pointing up. Putting the needle through a tissue, push the plunger until it reaches the desired volume to be injected.

Wipe off the needle with the tissue to remove any liquid on the outside of the syringe. Draw some air into the syringe. Hold the syringe with both hands. One hand is used to guide the needle and prevent bending while going through the septum. Push the barrel with the other hand while the thumb prevents the plunger from being pushed up by the gas pressure. Insert the needle through the septum and as far into the injection port as possible, depress the plunger, wait a second or two, then withdraw the needle (keeping the plunger depressed) as rapidly and smoothly as possible.

4   Analysis of Sterols from Microorganisms

4.1   Extraction of Sterols

Reagents   a) Ethanolic KOH: dissolve 15g of KOH in 100ml of 80% ethanol
b) Heptane
c) Chloroform

Procedure   1. Grow the organisms to the desired growth phase in a chemically defined medium.

2. Harvest the cells by centrifugation at 4000rpm for 15min then wash the cells three times with distilled water.

3. To each 2.0g wet weight of intact cells add 100ml of ethanolic KOH (a).

4. Extract the sterols according to the method of Fryberg et al. (1974) as follows:

Extraction   1. Transfer the suspension into a pre-weighed round bottom flask (RBF) and connect the condenser to the RBF.

2. Reflux the cell suspension in the RBF for 3 h under N_2, keeping the temperature at 60–70 °C and avoid boiling.

3. Filter the solvent using a pre-weighed filter paper and retain the filtrate.

4. Place the filter paper and the cells back into the RBF and dry the RBF, filter paper and cells by placing in an oven at 70 °C for overnight.

5. Weigh the RBF, the filter paper and cells, then subtract from the weight of the RBF plus the paper only. This will give the weight of the cells minus sterols.

6. To the filtrate obtained in step (3) add an equal amount of distilled water and then transfer to a separatory funnel (size 1 liter).

7. Add 50 ml of heptane to the separatory funnel and shake gently with pressure release every 3 s, then let it stand at room temperature for 10 min. The mixture will separate into two layers:

   - The upper layer contains heptane and the extracted sterols.

   - The lower layer is the aqueous phase containing saponified lipids and impurities.

8. Collect the upper layer and repeat step (7) three times on the aqueous phase.

9. Combine the upper layer from the four extractions and add anhydrous Na_2SO_4 to ensure that the heptane phase is water-free.

10. Filter the organic phase using filter paper then transfer the filtrate to a pre-weighed RBF.

11. Evaporate the heptane *in vacuo* using a rotary evaporator (temperature of the water bath should be between 70 and 80 °C).

12. Weigh the RBF after leaving in a desiccator containing anhydrous $CaCl_2$ for at least 30 min. Subtract the weight of the RBF from the weight of the RBF plus sterols to determine the amount of the extracted sterols.

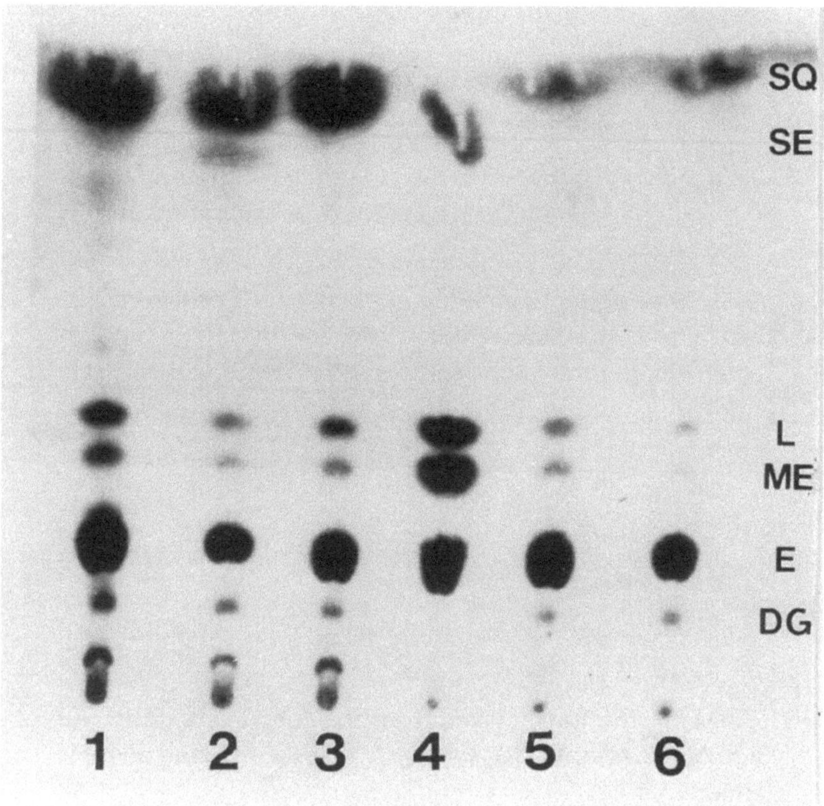

Fig. 9. Typical TLC plate showing sterols extracted from *Candida albicans*. The adsorbent was silica gel G; The solvent was petroleum ether (40–60 °C):diethyl ether (3:1, v/v). The sterols were visualised by charring. *1–6* Different sterol samples extracted from *C. albicans*; *SQ* squalene; *SE* steryl esters; *L* 4,4-dimethyl sterol (lanosterol); *ME* 4-methylsterol; *E* 4-4-desmethyl sterol (ergosterol); *DG* diacylglycerols

13. Dissolve each 10 mg of sterol in 1 ml of the chloroform and transfer into a vial.

14. Store the extracted sterols under N_2 atmosphere at $-20\,°C$.

4.2   Chromatography of Sterols

Thin layer chromatography. Sterols can be fractionated on TLC plates coated with silica gel G, using the solvent of petroleum ether: diethylether (75:25, v/v) at 40–60 °C. This solvent system should separate the sterols into squalene (the hydrocarbon precursor for sterol synthesis), steryl esters, 4,4-dimethyl sterol (e.g. lanosterol), 4-methyl sterol, 4,4-desmethyl sterol (e.g. ergosterol; (Fig. 9).

Gas liquid chromatography. Sterols can be analysed by GLC after derivatising them to the volatile form. One of the easiest and most convenient method is derivatising the sterols to trimethylsilyl derivatives (TMS) using the following method:

- Hexamethyl disilazane
- Trimethyl chlorosilane
- Chloroform
- Carbon disulphide (CS_2)

Reagents

1. Transfer 1 mg of total sterol dissolved in chloroform to a small glass – stoppered RBF (size 25 ml).

2. Evaporate the chloroform using N_2 then add 50 µl of hexamethyl disilazane and shake the flask gently and briefly.

3. Add 50 µl of 10% trimethyl chlorosilane (prepared in chloroform) to the reaction mixture and shake briefly then seal the RBF using a quickfit stopper that is greased with high vacuum grease.

4. Incubate the reaction mixture at room temperature for at least 4 h in a desiccator containing $CaCl_2$.

Gas liquid chromatography

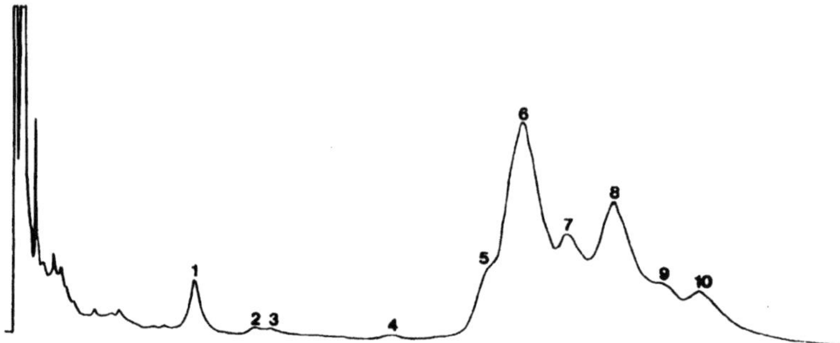

Fig. 10. Gas-liquid chromatograms on 3% SP-2100 of TMS derivatives of total sterols from *Candida albicans*. *1* Squalene; *2–3* breakdown products of ergosterol; *4* calciferol; *5* zymosterol; *6* ergosterol; *7* 4,14-dimethylzymosterol; *8* obtusifoliol; *9* lanosterol; *10* 24-methylenedihydrolanosterol

5. Remove the reagents by evaporating under N_2, then add 50 µl of carbon disulphide to dissolve the reaction mixture.

6. Using a gas chromatography syringe, inject 0.5 µl of the CS_2 dissolved mixture into a SP-2100 column (3% on 100/120 gaschrome Q; one can also use other non-polar stationary phases such as OV-1 or SE-30).

7. The GC should be adjusted to the following conditions:

   – Oven or column temperature = 230 °C

   – Injector temperature = 250 °C

   – Detector (FID) temperature = 300 °C

   – Carrier gas nitrogen or helium (30 ml/min)

8. Individual sterols (Fig. 10) can be identified either by comparing the retention time of the separated sterols with retention times of commercially available sterol standards or by comparing the relative retention time to ergosterol with available data from the literature (Vandenheuvel and Court 1968).

References

Dittmer JC, Lester RL (1964) A simple specific spray for the detection of phospholipids on thin layer chromatography. J Lipid Res 5:126–127

Fryberg M, Oehlschlager AC, Unrau AM (1974) Sterol biosynthesis in antibiotic-resistant yeast: Strains Arch Biochem Biophys 160:83–89

Vandenheuvel FA, Court AS (1968) Reference high efficiency non-polar packed column for the gas-liquid chromatography of nanogram amounts of sterols. Part I: retention time data. J Chromatogr 38:439–442

Wagner H, Lissau A, Holzi J, Horammer L (1962) The incorporation of ^{32}P into inositolphosphatides of rat brain. J Lipid Res 3:177–180

Weete JD (1980) Lipid biochemistry of fungi and other organisms. Plenum, New York

Chapter V   Phase Transition of Membrane Lipids

TIBOR PÁLI[1], BÉLA NÉMET[2] AND MIKLÓS PESTI[3]

1   Measuring Phase Transition by Electron Spin Resonance

T. PÁLI AND M. PESTI

1.1   Background

Membranes are multicomponent dynamic systems that play an important role in a number of essential cell processes. The mobility of molecules in membrane bilayer is a critical factor in considering the efficiency of these biochemical processes taking place in, or related to, the membranes. Since the viscosity of membranes cannot be defined and measured in a macroscopic sense, it is common practice to incorporate probe molecules (labels) into the bilayer and to measure how their rotational and translational diffusions are restricted. In fact, the parameters that describe the rate and amplitude of the motion of label molecules can be used to characterise the fluidity and order of the membrane.

Due to its optimal time scale and good sensitivity, the spin-label electron spin resonance (ESR) spectroscopy is an ideal

[1] Institute of Biophysics, Biological Research Centre, Hungarian Academy of Sciences, Szeged, Hungary

[2] Department of Physics, Janus Pannonius University, H-7624 Pécs, Ifjúság u. 6., Hungary

[3] Group of Microbiology. Department of Botany, Janus Pannonius University. Pécs, Hungary

tool for studying membrane dynamics. Both the rotational and translational diffusion of membrane proteins and lipids can be measured either in natural or artificial membranes (for review see Marsh and Horváth 1989). In ESR spectroscopy resonance absorption can be detected between the Zeeman levels of the unpaired electron, usually of the N-O bond in the spin label molecule. The spin label molecules for membrane studies are stable nitroxide free radicals that are synthetic analogues of natural lipids (Marsh 1985). There are labels available that covalently bind to -SH groups of proteins. The spin labelled lipids are incorporated in the membrane just as their natural analogues. Use of lipids, that are labelled at different C atoms along the hydrocarbon chain, makes it possible to study the molecular mobility at various depths of the membrane, targeted at a resolution of 1–$1.3\,\text{Å}$, in the membrane (Páli et al. 1992; Bartucci et al. 1993).

In an ESR experiment, the sample is positioned in the centre of the microwave resonator located between the two magnet poles. At a fixed microwave frequency ($\sim 9.1\,\text{GHz}$), magnetic field is scanned and the first derivative of the absorption from the microwave radiation is recorded. The time window of conventional ESR spectroscopy, determined by the spectral anisotropy of the nitroxide group, ranges from 10^7–$10^{11}\,\text{Hz}$ and, therefore, this method is optimal for studying molecular motions in the membrane. The sensitivity is also remarkble: one can detect 10^{-8}–$10^{-6}\,\text{M}$ spin label molecules in 5–$50\,\mu\text{l}$ volume. Due to the interaction between electron and nuclear spins of ^{14}N, the spectrum consists of three lines (as shown in Fig. 1). The exact position and width of these hyperfine lines depend on, among other factors, the rate and amplitude of rotation of the spin label molecule (Berliner 1989).

As membrane phase transitions are manifested not only by changes in the structure but also by changes in both the amplitude and rate of motion of the membrane molecules, the spin label ESR method is well suited to study phase transitions in membranes. The characteristic spectral parameters (Fig. 1)

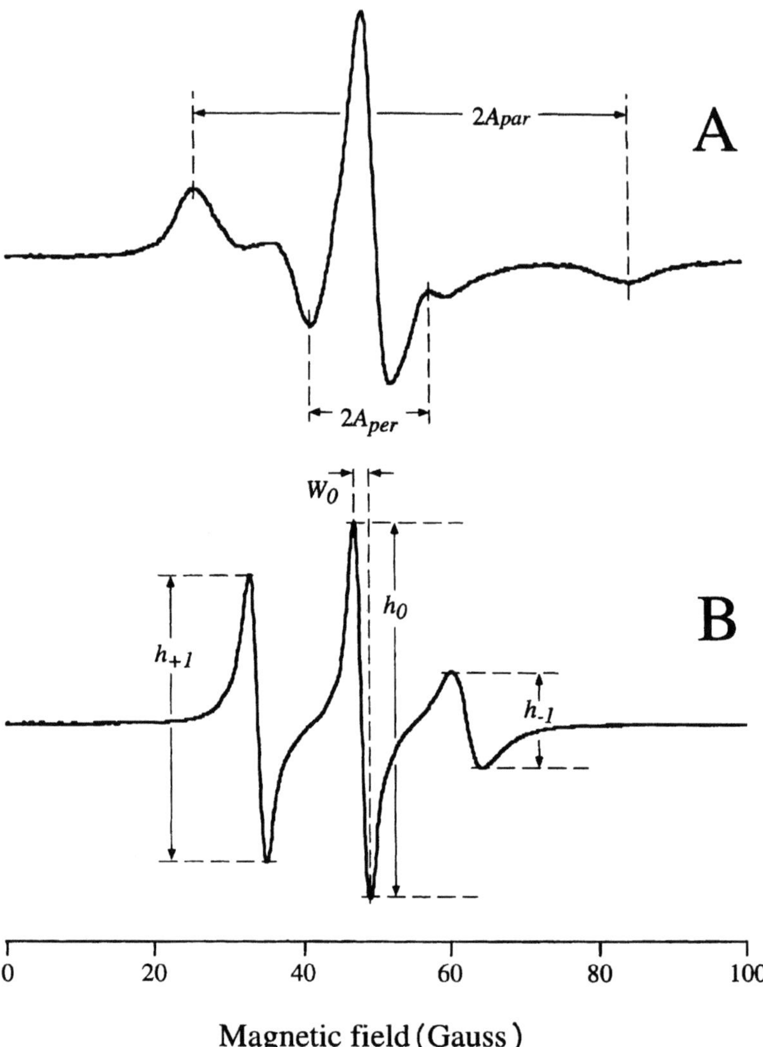

Magnetic field (Gauss)

Fig. 1A,B. Spectral parameters sensitive to rotational dynamics in the ESR spectra of positional isomers of phosphatidylcholine spin labelled in the sn-2 chain at the 5th (**A**) or 14th (**B**) carbon atom relative to the headgroup (5- and 14-PCSL, respectively), incorporated into small unilamellar vesicles of dimyristoyl phosphatidylcholine (DMPC). Spectra were recorded at 21 °C (**A**) and 36 °C (**B**). The spin label to lipid ratio was 1%. 14-PCSL monitors the central region of the bilayer where the motion is relatively isotropic, whereas 5-PCSL monitors the lipid headgroup region where the motion is more restricted. In addition, at 21 °C, the DMPC is still in the phase (**A**) where the motion is slow and the bilayer is more ordered relative to the liquid crystalline phase at 36 °C (**B**). See Eqs. (1) and (2) for a simple evaluation of the order parameter (S) and the rotational correlation time (τ_R) using the spectral parameters indicated in the figure

change abruptly at phase transition temperatures in pure lipid systems (Hubbell and McConnell 1971), as shown in Fig. 2A. In fact, alternative ESR approaches are applied to explore the molecular datails of phase transitions even below the main transition temperature (Páli et al. 1993).

1.2 Sample Preparation

1. In principle, there are membrane preparations with three levels of complexity for studying phase transitions: the original intact membrane structure, membrane fragments and artificial membranes made of the total lipid or a subclass of that, extracted from the original membrane system. While the former and more complex structures closely resemble the properties of the original membrane, the artificial membranes provide better defined phase behaviour and, therefore, are easier to interpret with often a questionable relevance to the original system.

   General procedure

   The membrane material is usually suspended in any buffer (no special consideration from the spin labelling point of view as long as it is harmless to the spin label) that should be close to the environment of the native membrane system. It is advisable to add the desired amount of spin label, dissolved in organic solvent, directly to the lipid solution.

   After drying the mixture under nitrogen flow and then under vacuum, hydrate them together to obtain membrane dispersions. Hydration usually requires vortexing the lipid-buffer mixture above the expected phase transition temperature for a few minutes. Alternatively, the preprepared aqueous dispersion of the lipid extract can be treated as natural membrane and labelled. See actual examples in protocols preparation of artificial membranes and preparation of natural membranes.

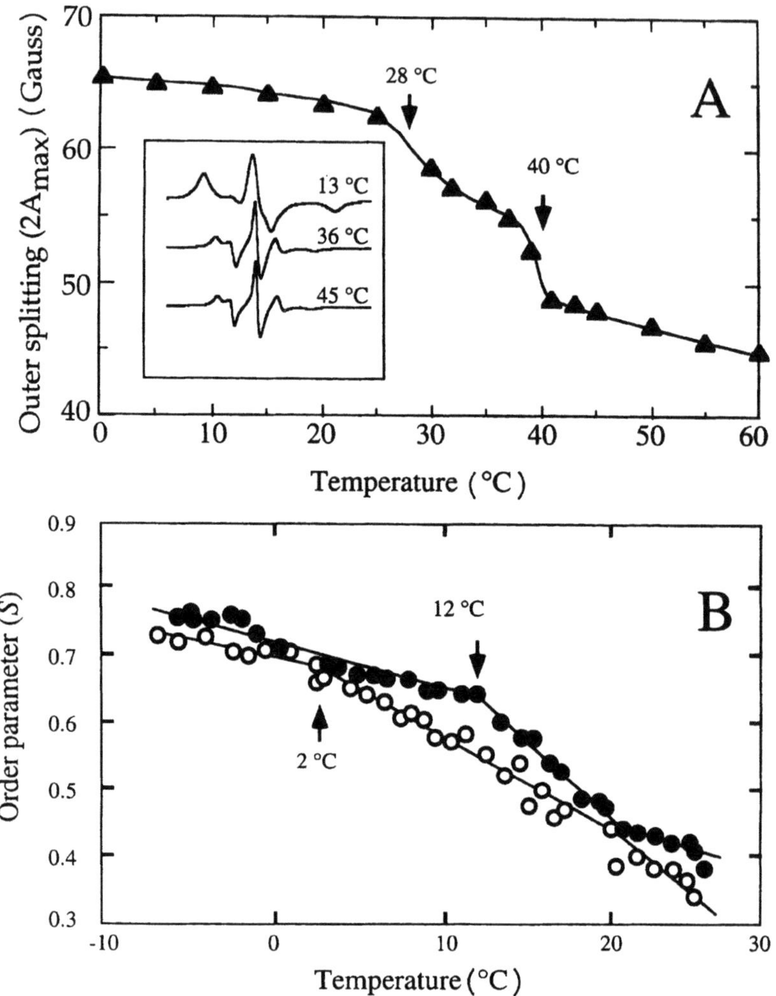

Fig. 2. A Outer splitting (2A$_{par}$, see Fig. 1 for definition) in dipalmitoyl phosphatidylcholine (DPPC) membranes labelled with 5-PCSL (0.5% relative to DPPC) as a function of temperature. The *insert* shows how spectral shape changes with temperature. **B** Temperature dependence of the effective order parameter [S. see Eq. (1) for evaluation] of the 5-doxylstearic acid spin label in membranes of an ergosterol-less (●) and ergosterol-producing mutant (○) of the yeast *Candida albicans*. *Arrows* indicate the temperature of the pre-and main transition of DPPC (**A**) and the transition of the yeast membranes (**B**).

2. Suspend the membrane preparation (whole cells or isolated membrane) in ~1 ml aqueous buffer (the suspension should correspond to 1 mg phospholipid). If there is no other consideration, the buffer can, for instance, be 20 mM HEPES, 10 mM EDTA, pH 7.4, or simply water.

   For a good spectrum one optimally needs 0.5–5 mg of phospholipid in the sample at a label:lipid ratio of 1:100. Any amount of phospholipid below 0.5 mg might result in a spectrum that is too weak, and excess of 5 mg of it cannot be loaded completely into the ESR capillary. It is therefore, advisable to estimate lipid phosphate (see Chap. IV, this Vol.) in order to quantitate the phospholipid content of the membrane preparation.

3. Make 1–10 mg/ml spin label stock solution of n-SASL [n-(4,4-dimethyloxazolidine-N-oxy) stearic acid] (Aldrich, Milwaukee, USA,) in ethanol. 5-SASL (n = 5) monitors the headgroup region of the bilayer and will measure the order parameter, whereas 16-SASL (n = 16), with the doxyl group located in the centre of a typical phospholipid membrane, detects the rotational mobility of the hydrocabon chains (see the motional parameters in Sect. 1.4).

   The quantity of the spin label should be estimated so that the resulting concentration of ethanol in the membrane suspension, and of the molar fraction of the label relative to the total lipid is by no means more than 2%.

4. Add the spin label solution to the membrane suspension during gentle shaking, vortexing or sonicating, depending on the sensitivity of the material to mechanical stress. Ethanol-mediated labelling has the advantage that it can be conveniently and quantitatively controlled. However, in order to avoid the use of ethanol (or any other organic solvent) the spin label should be dried at the bottom of a tube (under vacuum) and the membrane suspension should be layered onto it during vortexing.

Incubation time and temperature depend on the labelling efficiency; 1 h for n-SASL is typical. The spin label incorporates more readily in fluid membranes and, therefore, the temperature should be higher than the phase transition temperature but kept below 60 °C, as the spin label might degrade.

5. Remove ethanol and unincorporatred spin label by sedimentation or gradient centrifugation followed by resuspension in label-free buffer (about 1 mg phospholipid in 1 ml buffer). The conditions for this step depend strongly on the actual membrane, since the partial specific volume and possible surface charges strongly influence the sedimentation properties.

   Finally, concentrate the membrane suspension into 50–100 µl of buffer and load it into ESR capillaries (glass or quartz capillaries with internal diameter of 1 mm) or flat cells for bigger volumes.

! **Note.** Remove the supernatant or even enough material from the pellet if needed, after sedimenting the membrane suspension in the ESR capillary using a thin micropipette (with outer and inner diameter of 0.5–0.8 mm and 0.2–0.5 mm, respectively) to obtain a sample size of about 5–10 mm. Water absorbs microwave and thereby decreases the quality factor of the resonator, which results in a weaker signal.

Preparation of artificial membranes

Spin labelled dipalmitoyl phosphatidylcholine (DPPC) bilayers (for more details see Bartucci et al. 1993).

1. Dissolve 1–2 mg lipid in ethanol (1 mg/ml) and add n-PCSL (1-acyl-2-(n-(4,4-dimethyloxazolidine-N-oxy) stearoyl)-sn-glycero-3-phosphocholine) spin label (in this case 5-PCSL) to the lipid solution at a molar ratio of label:lipid of 1:200. Instead of n-PCSL one can also use n-SASL.

2. Evaporate the solvent and dry the sample under vacuum for at least 1 h.

3. Add 100 µl buffer (20 mM HEPES, pH 7.4) and vortex for 3 min at 50 °C where the DPPC is in the fluid phase.

4. Load the sample into a 100 µl capillary (with internal diameter of 1 mm) and pellet is at 3000 g for 10 min.

5. Remove the supernatant and leave no more than 5 mm pellet in the capillary.

6. After positioning the bottom 5 mm section of the capillary in the centre of the ESR resonator, set the sample temperature to about 30–40 °C below the expected main transition temperature (i.e. 41 °C for DPPC).

Spin labelling of the protoplast suspension of sterol mutants of the unicellular, petite-negative, human pathogenic yeast, *Candida albicans* (for more details see Pesti et al. 1985).

Preparation of natural membranes

1. Protoplasts can be isolated by any standard procedure (see Chap. II, this Vol.). Dilute the protoplast suspension (10^9 cells/ml) fivefold in 1 M mannitol.

2. Add 4 µl aliquot of a 5 mg/ml stock solution of 5-SASL in ethanol to 1 ml cell suspension.

3. To facilitate spin label incorporation, shake the mixture gently for 1 min.

4. To remove ethanol, sediment the protoplast suspension at low speed (1000 g for 5 min) and resuspend in 100 µl of 1 M mannitol.

5. Transfer the protoplast suspension into a micro flat cell (Scanlon, USA), centre it in the microwave resonator and set the starting temperature to −10 °C.

1.3 ESR Spectroscopy

Most of the commercial ESR spectrometers work in the X-band (9 GHz) region and are suitable for detecting stable nitroxide free radicals, such as membrane spin labels, as described below. Some of the well-known models are from Bruker (ESP series), Jeol (JES-PE series) and Varian (E-Line).

Spectroscopy

1. Spectrometer settings should be optimised for the actual spin labelled sample used. For the nitroxide spin labels mentioned here, a scan width of 100 Gauss centred around 3250 Gauss is typical. The centre depends on the type of the resonator and sample geometry.

   The modulation amplitude should be about 30% of the width of the sharpest peak. For smaller modulation amplitude the signal decreases linearly, for higher field modulation the signal becomes distorted. In practice, the modulation amplitude varies between 0.5–1.25 Gauss.

2. For low values, the signal amplitude is proportional to the square root of the microwave power, but at higher powers the signal saturates, i.e. it broadens and the amplitude increase becomes non-linear or even decreases. The saturation properties of the same label vary from sample to sample. Therefore one should set the microwave power such that it is still in the linear region (typically 5–10 mW).

3. A low pass filter is usually applied to improve the signal-to-noise ratio. The response time should be set to a value lower than 20–30% of the time needed for the recorder to pass the sharpest peak. This means that longer time constants require longer scans. Typical settings are 4 min, 0.25 s or 2 min, 0.128 s for the scan time and time constant, respectively.

4. Record a set of spectra with increasing temperature (2 °C steps) in a range ±15–30 °C around the phase transition.

In the example described in Section 1.2.2, the ESR spectra were recorded on an E-Line Century Series (Varian, USA) spectrometer. Instrument settings were: scan range, 3200–3300 Gauss; microwave power, 5 mW; modulation amplitude, 0.5

Gauss; scan time 4 min; time constant 0.25 s. In the example preparation of natural membranes in Section 1.2, spectra were recorded on a JES-PE-1X (JEOL, Japan) spectrometer with 4 min scans in the range fo 3200–3300 Gauss with a time constant of 0.25 s. Microwave power and modulation amplitude were 10 mW and 1 Gauss, respectively. Under the above conditions, due to unincorporated spin label no sharp isotropic triplet was observed.

1.4   Analysis

The ESR time window is classified into slow and fast motion regimes because there are different physical processes that determine the shape of the spectrum above or below of about 10^{-8} s. If the motion is slow, or fast but restricted, then the spectrum ranges over approximately 60 Gauss (Fig. 1A), depending on how much the spectral anisotropy is averaged out by the wobbling motion of the lipid at its long axis. In the fast motion regime, the angular amplitude of this motion is directly related to the order parameter S. The degree of the anisotropy, and therefore the order parameter, can be determined as follows:

1. Read the parameters A_{par} and A_{per} (shown in Fig. 1A) the hyperfine splitting of molecules that are oriented, parallel or perpendicular, with their long axis relative to the external magnetic field.

2. From these quantities measured in Gauss evaluate the molecular order parameter (S) using the relation:

$$S = \frac{A_{par} - A_{per}}{A_{zz} - \frac{1}{2}\left(A_{xx} + A_{yy}\right)}. \tag{1}$$

The constants in the denominator are the hyperfine splitting values of the spin label oriented in a single crystal host with its x, y and z molecular axes parallel to the external magnetic field. For a spin labelled stearic acid (n-SASL), the denominator is 27.55 Gauss. In real systems, S ranges between 0 and 1, where a higher-order parameter means a higher degree of ordering of the lipid chains.

The method can usually be applied for lipids labelled close to their headgroup (4- or 5-SASL, for instance) and for order parameters in range of S = 0.2–0.8. For more accurate values of S, a correction is required for the experimental A_{per} and for the polarity of the environment of the label. This correction, for single crystal values and/or other topics related to the order parameter, is discussed in detail in Berliner (1989).

3. In order to detect phase transition, or changes in the ordering state of the membrane, the order parameter is plotted as a function of temperature. In case A_{per} cannot be measured from the spectrum, simply plot A_{par} without evaluating S (see Fig. 2A).

   In the example Preparation of artificial membranes, Section 1.2, the outer splitting was determined by locating the outer extrema (see Fig. 1A) and by measuring the distance between them ($2A_{par}$). In order to detect phase transitions in DPPC, $2A_{par}$ was plotted as a function of temperature as shown in Fig. 2A. The transition at ~28°C is the pretransition between the gel and the rippled phase, whereas the main chain-melting transition appears at 40°C in this system.

   In the case of the yeast samples (for example, see Sect. 1.2), the order parameter was calculated using Eq. (1). The corrected value of A_{per} is determined using Gaffney's method (Berliner 1989). The phase transition of total lipid extracts are measured in lipid-water dispersion after hydrating the dry lipid/label mixture with 0.1 M NaCl + 5 mM $CaCl_2$ (pH 7).

Natural membranes are much more heterogeneous than those of hydrated lipids and they do not have sharp phase transitions. However, characteristic points in the temperature dependence of the order parameter or rotational correlation time could still be used to identify "phase transitions" as shown in Fig. 2B. The data presented are a clear indication that complex alterations in the lipid composition of the mutant cells are reflected at the plasma membrane level. These are evaluated by measuring the order parameter (for more details see Pesti et al. 1985).

If the motion is isotropic and fast, all three nitroxide lines **Rotational** are well resolved (Fig. 1B). The rate of random rotation is **correlation** quantitated by the rotational correlation time (τ_R): shorter **time** correlation time means faster motion, i.e. a more fluid membrane and vice versa. In the ESR spectrum, faster rotation results in narrower lines and the different sensitivity of the individual lines to rotation is used to calculate (τ_R) as follows:

1. Determine the peak to peak amplitudes and the central linewidth from the first derivative ESR spectrum as indicated in Fig. 1.B. The central linewidth (W_0) is measured in Gauss.

2. Evaluate the rotational correlation time (Kivelson 1960) as:

$$\tau_R = 0.65 \times W_0 \left(\sqrt{\frac{h_0}{h_{-1}}} - \sqrt{\frac{h_0}{h_{+1}}} \right). \qquad (2)$$

τ_R is obtained in ns. This method can be applied for lipids labelled close to the methyl end of their hydrocarbon chain (14- or 16-SASL) for rotational correlation times of about 10^{-11}–10^{-9} s.

3. Plot τ_R as a function of temperature. Even if τ_R cannot be evaluated, the central lineheight (h_0) can still be plotted because it changes steeply at the phase transition.

1.5 Tips and Tricks

- Reducing agents like ascorbate can decrease the spin label and quench the signal and, therefore, should be removed from the membrane suspension. For the same reason, active electron transport chains should also be blocked.
- The doxylstearic acid spin label partitions between membrane and aqueous buffer. Extensive washing of the labelled material can remove the label from the membrane. In addition, minimise the sample volume in the ESR capillary to reduce the disturbing signal originating from label molecules free in the buffer. If label is present in the buffer, special care must be taken to remove the mobile component from the spectrum when evaluating rotational correlation time (by subtraction) as it might significantly change the line height ratios (Fig. 1).
- Various spin labelled lipid analogues incorporate into membranes to the extent, where labelling level varies. To find optimal labelling conditions for quantitative analysis, use different amount of labels to test the labelling efficiency.
- Use of more than one spin label (5- and 14-SASL, for instance) might help to exclude artefacts caused by the sensitivity of the splitting or lineheights to special motional modes, and also to understand the nature of the phase transition. In order to determine the phase transition temperature more accurately, it is useful to record more spectra in the region where the transition is expected.
- To improve signal intensity, do not increase the labelling level, as spin-spin interactions will broaden the lines and distort the position, especially amplitudes, and the intensity would not increase as expected.

1.6 Alternatives

For labelling membranes, preference should be for spin labelled phospholipids since they are more analogous to natural ones than stearic acids; however, they are less readily available. The use of small (paramagnetic) molecules that partition between the membrane and the aqueous buffer like tetramethyl-piperidinyl-N-oxyl (TEMPO) is an alternative to spin labelled lipids or stearic acids to detect phase transitions in biomembranes. The partition coefficient depends on how densely the lipids are packed, and also on the phase of the membrane. In this case, the ratio of the two high-field spectral lines, belonging to the aqueous and membrane environments, is plotted as a function of temperature (Berliner 1989).

Several biochemical processes involve lateral movement of the components within the plane of the bilayer. Due to spin-spin interactions, the ESR spectrum is sensitive to the collision frequency of the spin label molecules incorporated into membranes and also to their lateral diffusion. A method for measuring lipid lateral diffusion coefficients with ESR is described in Páli and Horváth (1989).

In natural membranes or in lipid-protein complexes, a second spectral component is often observed, indicating that a certain fraction of the lipids has lower mobility. These immobilised lipids are involved in direct interaction with membrane proteins. With spectral subtractions or simulations it is possible to determine the fraction of the immobile lipids. By applying lipid labels with different headgroups and chain compositions, the lipid specificity of the membrane protein of interest can be explored (Marsh and Horváth 1989).

2 Measuring Phase Transition by Fluorescence

B. Nemét and M. Pesti

2.1 Background

The plasma membrane of living cells is a multicomponent system. To a certain extent, it is able to react to environmental effects (temperature, compounds, ions etc.), maintaining the homeostasis of the cell (Shinitzky 1984; Martonosi 1985). The phases, and the phase transitions of membranes, can be described by several characteristics (of the thermodynamic and physical characteristics of the membrane). These transitions can be suddenly induced by the change of temperature over a range. The phase transition temperature is one of the important characteristics of all experimental parameters that can be determined. The phase transitions are genereally of great importance in the field of membrane studies, as they are related to a large number of phenomena.

Membrane studies, based on the phenomenon of photoluminescence, have been carried out for more than a decade (Badley 1976; Dale 1983; Lakowicz 1985). Among various photoluminescent characteristics, the measurement of fluorescence anisotropy is commonly used for the description of phase transition (van der Meer et al. 1986; van Hoek et al. 1987; Kawski 1993). The steady-state fluorescence anisotropy measurements have been, and still are, used (van der Meer et al. 1986; Ben-Yashar and Barenholz 1989; Lakos et al. 1990; Schuler et al. 1990; Park and Huang 1992; Ansari et al. 1993); while time-resolved fluorescence aniosotropy kinetics examinations have become widespread only in the last decade (Kinoshita et al. 1982; Visser et al. 1985; van Hoek et al. 1987; Jameson et al. 1989; Honda et al. 1991; Behan et al. 1992; Yao et al. (1992). In addition, the method of fluorescence quenching (Henry-Toulme et al. 1989) and of fluorescence energy transfer are also applied in studies related to membrane dynamics (Henry-Toulme et al. 1989; Jona et al. 1990).

Membranes, in some cases, have their own fluorescent group of a relatively small size dispersed in the bilayer that can be used for fluorescence studies. Mostly, however, probe molecules, i.e. fluorophores like all-trans-1,6-diphenyl-1,3,5-hexatriene (DPH), 1-anilino-8-naphthalene sulfonate (ANS), dansyl chloride (DNS-C1), fluorescein-5′-isothiocyanate (FITC), seminaphoto-rhodafluor-1-acetoxymethylester (SNARF-1-AM), eosin, atebrin, aequorin, ethidium bromide (EtBr) etc., are introduced into the membrane and are used by monitoring their rotational diffusion that is hindered to different extents, depending on membrane lipid mobility (Badley 1976; Dale 1983; Visser et al. 1985; van Hoek et al. 1987; Ben-Yashar and Barenholz 1989; Jona et al. 1990; Honda et al. 1991; Sturtevant et al. 1991; Yao et al. 1992; Ansari et al. 1993).

The fluidity of the lipid bilayer influences many basic membrane functions. The fluidity strongly depends on the lipid composition of the bilayer (mainly on the sterols, phospholipids and sphingomyelin contents, and also on the degree of unsaturation of the phospholipid acyl chains), on the temperature and on the phase state of the membrane.

2.1.1 Steady-State Fluorescence Anisotropy (SSFA)

Steady-state fluorescence anisotropy is defined by an operational term as (Lakos et al. 1990):

$$r = \frac{\left(I_{VV} - I_{VV}^{S}\right) - \left(I_{VH} - I_{VH}^{S}\right)}{\left(I_{VV} - I_{VV}^{S}\right) + 2\left(I_{VH} - I_{VH}^{S}\right)}, \tag{1}$$

where the upper index S refers to the corresponding values of an unlabelled cell suspension and the I_{VV} and I_{VH} are the fluorescence intensities polarized parallel and perpendicular to the direction of polarization of the excitation beam.

The Perrin equation gives a relationsip between the measured anisotropy data and the molecular parameters (Lakowicz 1985; van der Meer et al. 1986; Schuler et al. 1990; Lakos et al. 1990):

$$\frac{r_{o}}{r} - 1 = C(r)\frac{\tau T}{\eta}, \tag{2}$$

where r and r_{o} are the measured and limiting anisotropies (the limit anisotropy value in a rigid medium is $r_{o} = 0.362$); η is the apparent "microviscosity" (the average microviscosity of the lipid region around the fluorophores); $(r_{o}/r - 1)^{-1}$ is the fluorescence anisotropy parameter; T is the absolute temperature; τ is the excited state lifetime; C(r) contains volume and shape factors for the fluorophore including associated solvent shell and any other factor contributing to curvature in the Perrin plot of the reciprocal of the emission anisotropy against the factor ($\tau T/\eta$).

2.1.2 Time-Resolved Fluorescence Anisotropy (TRFA)

The time-resolved polarized fluorescence provides detailed information on properties like mobility and local environment in cell membranes, nucleic acids, proteins, and on cell processes affected by various diseases, or modulated by drugs etc. (Kinoshita et al. 1982; van der Meer et al. 1986; Lakos et al. 1990; Honda et al. 1991).

Important characteristics of certain phases and phase transitions, such as the rotational correlation time (ϕ), the second rank order parameter ($<P_2>$) and the diffusion coefficient (D), can be obtained in the following way:

For calculation of the measured pseudo-fluorescence anisotropy decay, $r^*(t)$, and the pseudo-total fluorescence intensity, $s^*(t)$, we must measure the fluorescence decays, $I_V(t)$ and $I_H(t)$, when the excitation light pulses are vertically polarized, and the detection is through polarizers oriented vertically, $I_{VV}(t)$, and horizontally, $I_{VH}(t)$. G represents a factor to correct for depolarization due to detection system. The value of G ranges between 1.03 and 1.36 (Honda et al. 1991).

$$r^*(t) = \frac{I_{VV}(t) - I_{VH}(t)G}{I_{VV}(t) + 2I_{VH}(t)G} \tag{3a}$$

$$s^* = I_W(t) + 2I_{VH}(t)G. \tag{3b}$$

If g(t) is the apparatus response function of the exciting light, fluorescence anisotropy decay r(t) and total fluorescence intensity s(t) are calculated by the deconvolution of the following equations:

$$r^*(t) = \int g(t')r(t-t')dt \tag{4a}$$

$$s*(t) = \int g(t')s(t-t')dt. \tag{4b}$$

Both the anisotropy r(t) and the fluorescence s(t) are the sum of multiexponential decays:

$$r(t) = \frac{I_{VV}(t) - I_{VH}(t)}{I_{VV}(t) + 2I_{VH}(t)} = \sum \beta_i \exp\left(-\frac{\tau_i}{\phi_i}\right) \tag{5a}$$

$$s(t) = I_{VV}(t) + 2I_{VH}(t) = \sum \alpha_i \exp\left(-\frac{\tau_i}{\phi_i}\right), \tag{5b}$$

where τ_i is the fluorescence lifetime and ϕ_i is the rotational correlation time. Often r(t) is a sum of exponentials plus an r_∞ term;

$$r(t) = r_\infty + \sum \beta_i \exp\left(-\frac{\tau_i}{\phi_i}\right). \tag{6}$$

r_∞ represents the residual anisotropy at times longer ($t = \alpha \gg \tau$) than that compared to the fluorescence lifetime of the probe.

Several methods determine the decay parameters (α_i, τ_i, β_i, ϕ_i) from $s^*(t)$ and $r^*(t)$, such as the non-linear least-squares fit (van Hoek et al. 1987), global analysis (Flom and Fendler 1988), Laplace transform (van Hoek et al. 1987), etc.

The time scale of observation of rotational motion is dictated by the average of fluorescence or phosphorescence lifetime ($\tau_f \sim 10\,\text{ns}$, $\tau_{ph} \sim 100\,\mu\text{s}$) of the probe; 10% error in correlation times is involved for f between $0.1\,\tau$ and $100\,\tau$ (van Hoek et al. 1987).

The sum of the pre-exponential terms, $\Sigma\beta_i$ plus r_∞ represents the emission anisotropy at time zero.

$$r_0 = r_\infty + \sum \beta_i. \tag{7}$$

The following are special case:

- a single exponential decay plus constant ($\equiv r_\infty$)
- a double exponential decay plus a constant, or
- a double exponential decay.

A single exponential decay is a simple

$$r'(t) = (r_o - r_\infty)\exp\left(-\frac{\tau}{\phi}\right) + r_\infty,$$

(8)

where r_o is generally ~ 0.395 (van der Meer et al. 1986).

2.1.3 Order Parameter; Diffusion Coefficient

If the r_o and r_∞ values are known, the second rank order parameter $<P_2>$ (or $<S>$; from electron spin resonance measurements) is related by

$$\frac{r_\infty}{r_o} = <P_2>^2 .$$

(9)

The diffusion coefficient D can be calculated, as

$$D = \frac{1}{6} r_o \sum_i \left(\frac{\beta_i}{\phi_i}\right),$$

(10)

where D is the diffusion coefficient proportional to the size of the fluorescent particle

$$D = \frac{kT}{6V\eta},$$

(11)

where k is the Boltzmann constant, n is the solvent viscosity, V is the molecular volume and T is the absolute temperature.

2.1.4 Fluorescence Quenching (FQ) and Intramolecular Fluorescence Energy Transfer (IFET)

The fluorescence quenching and the energy transfer parameter (E) are calculated from the decrease of donor fluorescence intensity and the lifetime in the presence of acceptor (or quencher):

$$E = 1 - \frac{F_{DA}}{F_D} \quad \text{or} \quad E = 1 - \frac{\tau_{DA}}{\tau_D},$$

(12)

where F_{DA} and F_D or τ_{DA} and τ_D are the population averages of fluorescence intensities and lifetimes of the donor in the presence (F_{DA}, τ_{DA}) or the absence (F_D, τ_D) of the acceptor. The approximate distances, R, between the donor and acceptor moieties were estimated assuming random orientation using the following equation

$$R = R_o \left(\frac{1-E}{E} \right)^{1/6} ,$$ (13)

where R_o is the critical distance between donor (D) and acceptor (A), when E = 0.5 (E is the experimentally determined energy transfer efficiency), and it is given by

$$R_o(A) = 9.779 \times 10^{-3} \left(\frac{JK^2\Phi}{n^{-4}} \right)^{1/6} ,$$ (14)

where K^2 is a complex geometric factor that depends on the relative orientation of the transition dipoles of the donor and the acceptor. Assuming that donor and acceptor molecules rotate rapidly, as compared to the fluorescence lifetime of the donor, an average value of K^2 equal to 2/3 can be used. Φ is the fluorescence quantum yield of the donor, n is the refractive index of the medium (equal to 1.4), J is the normalized spectral overlap integral defined as:

$$J\left(M^{-1}\,m^{-1}\right) = \left(\frac{\int F_D(\lambda)\varepsilon(\lambda)\lambda^4 d\lambda}{\int F_D(\lambda)d(\lambda)} \right) ,$$ (15)

where $F_D(\lambda)$ is the donor fluorescence intensity and $\varepsilon_A(\lambda)$ ($M^{-1}cm^{-1}$) is the absorption coefficient at the wavelength λ (cm).

2.2 Objectives and Limitations

The study of membranes is very extensive, and fluorimetric methods can give only a rough outline of the applied membrane preparation methods and of the experimental results.

For this reason, the sample preparation methods are classified into three parts:

- membrane studies on different lamellar vesicles
- membrane studies on samples prepared from cells and
- membrane studies on living cells.

The experimental results presented are classified based on different spectrofluorimetric parameters used:

- results of SSFA measurements
- results of TRFA measurements and
- results of FQ and IFET measurements.

2.3   Materials and Procedures

Below are some typical methods of dye-labelled membrane preparations, e.g. frequently used solvents, buffers, reagents, critical parameters, optimal conditions, alternative procedures (for labelling by dyes), storage conditions and equipment for the fluorimetric measurements.

Preparations of unilamellar and multilamellar vesicles

Large unilamellar vesicles preparation is adopted from Schuler et al. (1990), see also Chapter IX, this Volume.

1. For large unilamellar vesicles preparation use 0.1 M NaCl buffered with 5 mM HEPES (pH 7.4) by reverse-phase evaporation from soybean PC, PG and sterol(s) in different molar ratios.

2. Introduce systematically in the assays, 1 mol % of butylated hydroxy-toluene to avoid the oxidation of phospholipid fatty acyl chains.

3. Introduce argon atmosphere in all steps involved in the preparation of vesicles.

4. Use chloroform and chloroform: methanol (98:2, v/v) respectively for preparation of sterol and PL stock solutions and store at −20 °C in sealed vials.

5. Label vesicles with 10^{-6} M DPH solution in tetrahydrofuran (concentration of solution was 10^{-3} M) at a final total lipid (PL + sterol) to probe molar ratio of 150–200:1.

6. Incubate suspensions for 30 min in dark at room temperature just prior to fluorescence measurements (Schuler et al. 1990).

Note. Steady-state fluorescence anisotropy was measured on an SLM 8000 spectrofluorimeter in the T-format. !

Multilamellar large vesicles (MLV) preparation is adopted from Ben-Yashar and Barenholz (1989).

1. Prepare the ML V in 50 mM KCL containing 15% sucrose. The presence of sucrose prevents precipitation of MLV.

2. Dissolve the probes [all-trans-1,6-diphenyl-1,3,5-hexatriene (DPH), 4-heptadecyl-7-hydroxycoumarin (HC), all-trans-9,11,13,15-parinaric acid (TPA)] in tetrahydrofuran (THF) and inject them into the MLV dispersion. Use a probe/lipid mole ratio of 1:1000–1:2000 for DPH and 1:250 for HC and TPA. The final concentration of THF should not exceed 0.1% of the aqueous solution.

3. Vortex the dispersion and incubate at 50 °C (above the DPPC gel to liquid crystalline phase transition) till fluorescence intensity reaches a plateau (usually not more than 1 h).

Note. Steady-state fluorescence anisotropy was measured with MPF-44 Perkin-Elmer spectrofluorimeter. !

Isolation of mitochondria and liposomes is adopted from Senault et al. (1990).

Preparations of membranes extracted from mitochondrial lipids

1. Use male weaning rats for the isolation of mitochondria and liposomes after decapitation of the animals.

2. Dissect the inter-scapular brown tissue rapidly, rinse and homogenise in ice-cold sucrose buffer (250 mM sucrose; 5 mM TES, pH 7.2).

3. Suspend the mitochondria in 250 mM sucrose; 5 mM TES (pH 7.2). Details of mitochondria preparation from rat brown tissue can be obtained from Senault et al. (1990).

4. After extracting the mitochondrial lipids (as described in Chap. III, this Vol.), the extract is evaporated under vacuum.

5. Hydrate the uniform layer of mitochondrial lipids just before physical measurements with 0.25 M sucrose, 5 mM K-TES (pH 7.2), 2 mM EDTA, 5 μM rotenone.

6. Vortex five times for 30 s at room temperature.

7. Label mitochondria and liposomes with a solution of DPH 1.25×10^{-4} M in THF. When the molar ratio of DPH to lipids is 1:200, dilute the mitochondrial samples (0.125 mg protein/ml) and incubate with DPH for 30 min at room temperature.

8. Use sucrose-based medium, which has the advantage of maintaining mitochondrial integrity, during measurements.

! **Note.** Flourescence anisotropy measurements were done with a T-format SLM 8000 spectrofluorimeter at 25 °C.

Isolation of plasma membrane from yeast cells

Spheroplast preparation from exponentially grown yeast cells is adopted from Ansari et al. (1993).

1. Suspend the rigorously washed *Candida albicans* cells (1 g wet wt) in 10 ml of 100 mM Tris-Cl (pH 8.8) containing 50 mM 2-mercaptoethanol (2-ME).

2. Incubate the suspension at 30 °C in a rotary shaker bath for 30 min.

3. Remove the 2-ME and resuspend the cells in a solution containing 20 mM $MgSO_4$ and 0.6 M sorbitol.

4. Add Zymolyase 20 T (20 units mg^{-1}) to the spheroplast suspension (8 mg dry wt) to remove the cell wall. Incubate this suspension once again at 30 °C for 3 h with gentle shaking. Check the formation of spheroplasts by their lysis as described in Chapter II, this Volume.

5. Prepare DPH (2 mM) in THF for labelling of the spheroplasts. Add 100 µl of DPH to 50 ml of rapidly stirring phosphate buffer (10 mM, pH 6.8). Remove the excess of THF by flushing with nitrogen.

6. Wash the spheroplasts with phosphate buffer (20 mM, pH 6.0) containing 10 mM $MgSO_4$ and 0.6 M sorbitol and incubate with 2 µM DPH for 60 min at 30 °C.

7. Spheroplasts can also be labelled with 1-anilino-8-naphthalene sulfonate (ANS). In this case, spheroplast samples containing 100 µg/ml protein, 50 mM Tris-acetate (pH 7.2) and 33.3 µM ANS respectively, were used for fluorescence polarisation studies.

8. Fluorescence polarisation studies can be carried out on these preparations.

Note. Fluorescence polarisation measurements were done in ! SFM-25, Kontron Instruments, Switzerland.

Isolation of plasma membranes from Balb/C cells

Preparation of dye-labelled Balb/C mouse thymocytes is adopted from Lakos et al. (1990).

1. Suspend Balb/C mouse thymocytes (5×10^7 cells/ml) in normal phosphate-buffered saline (PBS, pH 7.4).

2. For labelling, 4×10^{-3} mM DPH (in THF) is diluted 1000-fold by injecting into a vigorously stirred normal PBS at room temperature. Incubate for 1 h in dark to avoid unfavourable photoisomerization of DPH and to evaporate

THF. Mix the DPH solution 10:1 (v/v) with the cell suspension (5×10^7 cell/ml) and incubate for 30 min at 37 °C in dark.

3. Other dyes can also be used for labelling the cells.

! **Note.** SSFA measurements were carried out in a Hitachi MPF-4 spectrofluorimeter.

2.4 Results

Some examples are discussed, on one hand relating to the measurement of different fluorescence quantities (steady-state fluorescence anisotropy, time-resolved fluorescence anisotropy, fluorescence quenching, fluorescence intramolecular energy transfer), and on the other hand relating to the composition of the membrane samples examined.

2.4.1 SSFA Investigations

Schuler et al. (1990) compared the typical plant sterols (sitosterol, stigmasterol and campesterol) with respect to their ability to regulate membrane fluidity of soyabean PC vesicles. Fluidity changes were monitored by the SSFA with DPH as a probe and assigned to a measure of the acyl chain orientational order. Sitosterol and campesterol appear to be the most efficient sterols in ordering the acyl chains of soybean lecithin bilayers.

Ben-Yashar and Barenholz (1989) compared cholesterol-phospholipid and cholest-4-en-3-one-phospholipid interactions by their effect on thermotropic behaviour of DPPC bilayers. This was determined by the temperature-dependent SSFA of three fluorophores; DPH, HC and TPA. To monitor different lateral and vertical locations of the lipid bilayers, the fluorophores were used; DPH and HC average laterally the properties of the hydrophobic and head group regions of the bilayer, respectively, while TPA distribution is determined by the lateral organization of the bilayer. The data show that both the steroids diminish the pre-transition and behave as "averagers", bordering the main gel to liquid crystalline phase transition through ordering of acyl chains in the liquid crystalline state and disordering of them in the gel state.

Ansari et al. (1993) used polyene-resistant ergosterol-less mutants of *Candida albicans* to study with SSFA the role of various accumulated sterol intermediates on amino acid transport. The uptake of a number of amino acids was reduced in erg mutant strains as compared with the wild-type 33ERG[+]. In order to ascertain the relationship between membrane permeability and physical state of membrane, the fluorescence polarisation measurements were done with the erg mutants. It was observed that the depletion of ergosterol in erg mutants did not result in increased membrane fluidity; instead, the accumulated biosynthetic precursors led to an even higher level of increase in membrane rigidity. The decrease in amino acid transport was due to an increase in membrane order.

Lakos et al. (1990) monitored different structural regions of the bilayer and the lipid fluidity using three fluorescent lipid probes: TMA-DPH, 12-AS and DPH. The SSFA of these probes was measured as a function of temperature at two values of transmembrane potential. DPH was proved to be dependent on membrane potential in high temperature range (above 28 °C), while no such dependence could be observed for DPH below this temperature range. For TMA-DPH and 12-AS, the dependence was observed between 20 and 37 °C. These data suggest that changes in transmembrane potential are accompanied by local alteration in membrnae lipid dynamics and/or structure.

2.4.2 TRFA Investigations

Behan et al. (1992) used SSFA and TRFA of DPH fluorescence to compare the hydrocarbon order of brain myelin membranes from a shallow water (plaice) and two deep-sea fish species (*Coryphenoides rupestris* and *Coryphenoides armatus*). Time-resolved measurements allowed the separate determination of the rate of probe motion from the amplitude of that motion. Anisotropy decays were analysed in terms of two correlation times and a constant (r_∞). The r_∞ and $<P_2>$ order parameter for all species increased with hydrostatic pressure.

Yao et al. (1992) have made a comparative study of rotational relaxations of an eosin-protein complex, using delayed fluorescence and phosphorescence as well as the anisotropy parameters of phosphorescence and delayed fluorescence from eosin-bound bovine serum albumin. The results indicate that the measurement of delayed fluorescence is often the preferable option for investigating the rotational diffusion of the eosin bound macromolecules.

Senault et al. (1990), using DPH as a probe, estimated the parameters of membrane fluidity by SSFA (r_s) and by TRFA (order parameter S). Cold acclimation induced a decrease of PC/PE ratio, an increase of the total fatty acid unsaturating index (T.U.). Essential fatty acids (EFA) deficiency had the same effect as cold on the PC/PE ratio, but decreased T.U. Cold adaptation induced a larger decrease of S than of r_s, whereas EFA deficiency only increased r_s and did not modify S.

2.4.3 FQ Investigations

Park and Huang (1992) studied the interaction of synthetic glycophospholipids with phospholipid bilayer membranes by fluorescence quenching method. A series of glycophospholipids, synthesized by coupling mono-, di-, or tri-saccharides to dioleoylphosphatidylethanolamine (DOPE) by reductive amination, were used to investigate the interaction of glycophospholipids with phospholipid bilayer membranes. The hydration intensity of the glycophospholipid was directly related to its ability to stabilize the DOPE bilayer phase for vesicle formation. Glycophospholipids also reduced the transition temperature from gel to liquid-crystalline phase (T_m) of DPPC bilayers. Incorporation of N-neuraminylactosyl-dioleoyl-phosphatidylethanolamine (NANL-DOPE) induced a decrease of membrane fluidity of DPPC bilayers in the gel phase, while other glycophospholipids had no effect. Furthermore, a low level of NANL-DOPE increased the transition temperature (T_H) from the liquid-crystalline to the hexagonal phase of dielaidoylphosphatidylethanolamine (DEPE) bilayers.

2.4.4   IFET Investigations

Henry-Toulme et al. (1989) studied the interactions of the polyene antibiotics [Amphotericin B (AmB), N-(1-deoxyD-fructos-1-yl) amphotericin B (N-Fru-AmB)] with murine thymocytes by fluorescence energy transfer from TMA-DPH to polyenes. Polyene location was investigated by comparing the energy transfer efficiency obtained with TMA-DPH and with the parental compound DPH. While AmB binds to plasma membrane, as well as to intracellular structures, N-Fru-AmB seems to accumulate into the cell and bind to intracellular membrane structures. The effect of the two polyenes on membrane fluidity was studied by SSFA.

Jona et al. (1990) investigated the temperature dependence of fluorescence polarization and Förster-type resonance energy transfer (FRET) in the Ca^{2+}-ATPase of sarcoplamic reticulum, using protein tryptophan fluorescence and site-specific fluorescence indicators [5-{2-((iodoacetyl) amino) ethyl} amino naphthalene-1-sulfonic acid] (IAEDANS), fluorescein-5′-isothiocyanate (FITC), 2′,3′-O-(2,3,4-trinitrophenyl) adenosine monophosphate (TNP-AMP) and probes (lanthanides: Pr^{3+}, Nd^{3+}). The thermally induced structural fluctuations increased the energy transfer, which in effect indicated the existence of a relatively flexible structure in the region of the ATPase molecule that linked the IAEDANS to the FITC site.

2.5   Trends in Spectrofluorimetric Studies

Microspectrofluorimeter using tuned pulsed dye laser is able to investigate intracellular of isolated living cell, when the excited volume is less than $5\,\mu m^3$. Sturtevent et al. (1991) determined the intracellular pH of *Candida albicans* cells by laser micro-spectrofluorimeter with a pH-sensitive fluorescence probe, seminaphotorhodafluor-1 (SNARF-1). They confirmed that *Candida albicans* is able to maintain its intracellular pH (which is dependent on the morphology of the cell) over a wide range of external pH.

Destrampe and Hieftje (1993) developed a rapid-scanning spectrofluorimeter for measuring a series of multidimensional excitation-emission polarisation contour-plots. Time-consuming of an 80 nm × 80 nm block measurement is 20 s. Gough and Taylor (1993) used fluorescence anisotropy imaging microscopy to release maps of calmodulin binding during cellular contraction SNARF-1 and locomotion.

References

Ansari S, Gupta P, Mohanty SK, Prasad R (1993) The uptake of amino acids by *erg* mutants of *Candida albicans*. J Med Vet Mycol 31:1–10

Badley RA (1976) Fluorescent probing of dynamic and molecular organization of biological membranes. In: Wehry EL (ed) Modern Fluorescence Spectroscopy, Vol 2. Plenum Press, New York, pp 91–168

Bartucci R, Páli T, Marsh D (1993) Lipid chain motion in an interdigitated gel phase: Conventional and saturation transfer ESR of spin labelled lipids in dipalmitoylphosphatidyl choline-glycerol dispersions. Biochemistry 32(1):274–281

Behan MK, Macdonald AG, Jones GR, Cossin AR (1992) Homeoviscous adaptation under pressure: the pressure dependence of membrane order in brain myelin membranes of deep-sea fish. Biochim Biophys Acta 1103:317–323

Ben-Yashar V, Barenholz Y (1989) The interaction of cholesterol and cholest-4-en-3-one with dipalmitoyl phosphatidylcholine. Comparison based on the use of three fluorophores. Biochim Biophys Acta 985:271–278

Berliner LJ (ed) (1989) Spin labelling: theory and applications. Academic Press, New York

Dale RE (1983) Membrane structure and dynamics by fluorescence probe depolarization kinetics. Time-resolved fluorescence spectrscopy in biochemistry and biology. NATO ASI Ser 69:555–605

Destrampe KA, Hieftje GM (1993) New instrumentation for use in excitation-emission fluorescence polarization measurements. Appl Spectrosc 47:1548–1554

Flom SR, Fendler JH (1988) Global analysis of fluorescence depolarization experiments. J Phys Chem 92:5908–5913

Gough AH, Taylor DL (1993) Fluorescence anisotropy imaging microscopy maps calmodulin binding during cellular contraction and locomotion. J Cell Biol 121:1095–1107

Henry-Toulme N, Semon M, Bolard J (1989) Interaction of amphotericin B and its N-fructosyl derivative with murine thymocytes: a comparative study using fluorescent membrane probes. Biochim Biophys Acta 982:245–252

Honda S, Morii H, Ohashi S, Vedaira H (1991) Fluctuation and rotation of human growth hormone-releasing factor in the presence and the absence of phospholipid bilayer analyzed by time-resolved fluorescence depolarization. Biochim Biophys Acta 1068:81–86

Hubbell WL, McConnell HM (1971) Molecular motion in spin-labelled phospholipids and membranes. J Am Chem Soc 93:314–326

Jameson DM, Thomas V, DeMing Zhou (1989) Time-resolved fluorescence studies on NADH bound to mitochondrial malate dehydrogenase. Biochim Biophys Acta 1989:187–190

Jona I, Matkó J, Martonosi A (1990) Structural dynamics of the Ca^{2+}-ATPase of sarcoplasmic reticulum. Temperature profiles of fluorescence polarization and intramolecular energy transfer. Biochim Biophys Acta 1028:183–199

Kawski A (1993) Fluorescence anisotropy – theory and applications of rotational depolarization. Crit Rey Anal Chem 23:459–530

Kinoshita S, Tanaka I, Kushida T, Kioshita Y, Kimura S (1982) Time-dependent fluorescence depolarization of fluorescein in rat thymus lymphocytes. J Phys Soc Jpn 51:598–604

Kivelson D (1960) Theory of ESR line-widths of free radicals. J Chem Phys 33:1094–1107

Lakos Z, Somogyi B, Balázs M, Matkó J, Damjanovich S (1990) The effect of transmembrane potential on the dynamic behaviour of cell membranes. Biochim Biophys Acta 1023:41–46

Lakowicz JR (1985) Principles of fluorescence spectroscopy. Plenum Press, New York

Marsh D (1985) ESR probes for structure and dynamics of membranes. In: Bayley PM, Dale RE (eds) Spectroscopy and dynamics of molecular biological systems. Academic Press, London, pp 209–258

Marsh D, Horváth LI (1989) Spin-Label studies of the structure and dynamics of lipids and proteins in membranes. In: Hoff AJ (ed) Advanced EPR applications in biology and biochemistry. Elsevier, Amsterdam pp 707–752

Martonosi A (ed) (1985) The enzymes of biological membranes, Vol 3. Plenum Press, New York

Páli T, Horváth LI (1989) Restricted lateral diffusion of acidic lipids in phospholipid vesicles aggregated by myelin basic protein. Biochim Biophys Acta 984:128–134

Páli T, Bartucci R, Horváth LI, Marsh D (1992) Distance measurements using paramagnetic ion-induced relaxation in the saturation transfer electron spin resonance of spin labelled biomolecules. Application to phospholipid bilayers and interdigitated gel phases. Biophys J 61:1595–1602

Páli T, Bartucci R, Horváth LI, Marsh D (1993) Kinetics and dynamics of annealing during the sub-gel phase formation in phospholipid bilayers: saturation transfer electron spin resonance study. Biophys J 64:1781–1788

Park VS, Huang L (1992) Interaction of synthetic glycophospholipids with phospholipid bilayer membranes. Biochim Biophys Acta 1112:251–258

Pesti M, Horváth LI, Vígh L, Farkas T (1985) Lipid content and ESR determination of plasma membrane order parameter in Candida albicans sterol mutants. Acta Microbiol Hung 32(4):305–313

Schuler I, Duportail G, Glasser N, Benveniste P, Hartmann M (1990) Soybean phosphatidylcholine vesicle containing plant sterols: a fluorescence anisotropy study. Biochim Biophys Acta 1028:82–88

Senault C, Yazbeck J, Goubern M, Portet R, Vincent M, Gallay J (1990) Relation between membrane phospholipid composition, fluidity and function in mitochondria of rat brown adipose tissue. Effect of thermal adaptation and essential fatty acid deficiency. Biochim Biophys Acta 1023:283–289

Shinitzky M (ed) (1984) Physiology of membrane fluidity, Vol 1. CRC Press, Boca Raton, pp 131–172

Sturtevant J, Seksek O, Bolard J (1991) Intracellular pH and morphological transition in Candida albicans. Laser microspectrofluorimetry studies on isolated cells. J Mycol Med 1:208–211

van der Meer BW, van Hoeven RP, Bitterswijk WJ (1986) Steady-state fluorescence polarization data in membranes. Resolution into physical parameters by an extended Perrin equation for restricted rotation of fluorophores. Biochim Biophys Acta 854:38–44

van Hoek A, Vos K, Visser AJWG (1987) Ultrasensitive time-resolved polarized fluorescence spectroscopy as a tool in biology and medicine. IEEE J Quantum Electron 23:1812–1820

Visser AJWG, Ykema T, van Hoek A, O'Kane GJ, Lee J (1985) Determination of rotational correlation times from deconvoluted fluorescence anisotropy decay curves. Demonstration with 6,7-dimethyl-8-

ribityllumazine and lumazine protein from Photobacterium leiognathi as fluorescent indicators. Biochemistry 24:1489–1496

Yao J, Mictay D, Rogers AJ, Quinn PJ (1992) A comparative study of rotational relaxation of an eosin-protein complex using delayed fluorescence and phosphorescence. J Mod Opt 39:2363–2373

Chapter VI Lipid Asymmetry of Membranes

Vijay K. Kalra[1], C.M. Gupta[2], A. Zachowski[3]
and T. Pomorski[4]

Three sections of this chapter describing various approaches to study membrane lipid asymmetry are written by three different groups. Each section is preceded by a brief introduction to give a general background of the respective approach. These methods were initially worked out to study lipid asymmetry of RBC membrane, but these could well be adopted to other membrane systems.

1 Chemical Probes to Study Lipid Asymmetry

V.K. Kalra

1.1 Background

Phospholipids are the most abundant type of lipids in cell membranes of all eukaryotic and prokaryotic cells. Studies in the early 1970s revealed that these phospholipids are arranged as a bilayer in the cell membrane (Stoeckenius and Engelman 1969; Singer 1974). In order to determine the distribution of these phospholipid molecules in the lipid bilayer, both chemi-

[1] Department of Biochemistry, University of Southern California, School of Medicine, Los Angeles, California 90033, USA
[2] Institute of Microbial Technology, Chandigarh-160014, India
[3] Institu de Biologie Physico-Chimique: rue Pierre et Marie Curie, 75005 Paris, France
[4] Humboldt Universität zu Berlin, Mathematisch-Naturwissenschaftliche Fakultät 1, Institut für Biologie/Biophysik: Invalidenstrasse 43, 10115, Berlin, Germany

cal and enzymatic probes have been utilised (Bretscher 1972; Gordesky and Marinetti 1973; Verkleij et al. 1973). Most of the earlier studies on the topology of phospholipids in the membranes were conducted on human red blood cells as these are devoid of intracellular organelles. The presence of intracellular organelles can complicate the interpretation of the data, as phospholipids, derived from the intracellular organelles, can undergo labelling and modification by chemical probes if conditions for labelling are such that probe becomes permeant to the membrane (Marinetti and Crain 1978).

It was earlier thought that the function of phospholipids was to maintain the structure and fluidity of the membrane. A number of studies have shown that phospholipids present in cell membranes of red blood cells, platelets and some mammalian and bacterial cells (so far examined), display asymmetric distribution across the lipid bilayer (Marinetti and Crain 1978; Op den Kamp 1979; Rawyler et al. 1984; Fadok et al. 1992). In normal red blood cells, phosphatidylcholine (PC) and sphingomyelin are primarily present in the external leaflet of the bilayer. Most of the phosphatidylethanolamine (PE) and all the phosphatidyl serine (PS) are localised in the inner cytoplasmic leaflet (Verkleij et al. 1973).

Studies by Seigneuret and Devaux (1984) have revealed that aminophospholipid translocase enzyme mediates the translocation of PS and PE from the outer to the inner leaflet of the bilayer in an ATP-dependent manner. The loss of asymmetry of phospholipids, specifically the appearance of a small percentage of total PS in the external leaflet of the bilayer of sickle red blood cells, has been shown to be due to a decrease in the aminophospholipid translocase activity in these cells (Blumefield et al. 1991). The presence of PS in the outer leaflet of the lipid bilayer of red blood cells, e.g. sickle RBC, has been reported to play a role in the adherence to endothelial cells (Schlegel et al. 1985; Kalra et al. 1990) and recognition by monocytes/macrophages (Tanaka and Schroit 1983; Schwartz et al. 1985; McEvoy et al. 1986). In addition, the role of specific phospholipids present in the membrane of RBC in chemically induced fusion has been demonstrated (Farooqui et al. 1987). These studies demonstrate the functional importance of phospholipids in biological membranes, and thus, non-permeant probes, both chemical and enzymatic, are expected to determine the orientation of phospholipids in the cell membrane.

Trinitrobenzene sulfonic acid (TNBS) and fluorescamine, which react with amino groups, have been utilised as vectorial agents to study the orientation of phospholipids in the membrane (Gordesky et al. 1975; Marinetti and Crain 1978; Rawyler et al. 1984; Franck et al. 1986; Farooqui et al. 1987; Wali et al. 1988). These reagents react primarily with amino groups of proteins and phospholipids and, under appropriate conditions (e.g. buffer, temperature, pH and length of incubation), can be utilised to label externally oriented phosphatidylserine and phosphatidyl-ethanolamine in the lipid bilayer of cell membrane. TNBS dissolved in bicarbonate buffer, pH 8.6, has been observed to react optimally with amino groups of externally oriented amino-phospholipids of red blood cell membranes when incubated at room temperature (22–23 °C; Gordesky et al. 1975). The optimum labelling takes 30 min to 2 h. The reaction of TNBS with the amino group of PE to form a trinitrophenylated-PE (TNP-PE) derivative is mostly complete within this time period. Any longer incubation period (3 h) results in penetration of the probe, leading to labelling of both aminophospholipids in the inner leaflet and haemoglobin. The probe, TNBS, becomes permeant and causes haemolysis of RBC when phosphate buffer is utilised or when a temperature of 37 °C and an incubation time of more than 3 h is utilised for labelling.

1.2 Materials and Methods

1.2.1 Labelling of Cells with TNBS

Materials • Trinitrobenzene sulfonic acid (Picryl sulfonic acid, TNBS; Sigma Chemical Co., St Louis, Mo)

- Buffer A. 120 mM sodium bicarbonate ($NaHCO_3$) containing 40 mM NaCl in distilled water, pH adjusted to 8.5 to 8.6 (bicarbonate-saline buffer)
- Cells. Blood (5–10 ml) collected in heparinised tubes or tissue cultured cells (1×10^7 cells washed with isotonic saline)

1. Prepare 100 ml of 2 mM TNBS dissolved in buffer A.

2. Centrifuge blood at 3000 rpm for 10 min. The plasma and buffy coat at the top of red cells is removed by Pasteur pipette. To the pellet add 10 ml of isotonic saline (145 mM NaCl) and centrifuge at 3000 rpm for 10 min. The supernatant is discarded and this procedure is repeated one more time. The pellet is suspended to a hematocrit of 25 to 30% in isotonic NaCl.

3. Buffer A (20 ml) is added to an aliquot of 0.5 ml of red blood cells in a 30 ml Corex tube. This sample will be used as a control.

4. Buffer A (20 ml) containing 2 mM TNBS is added to another aliquot of 0.5 ml of red blood cells.

5. Incubate the contents of both tubes for 30 min to 2 h at 22–23 °C. The pH of the suspension should remain between 8.5 to 8.6. If necessary, adjust the pH with either NaOH or dilute HCl. Incubate contents at 22–23 °C for 30 min to 2 h.

6. After incubation, centrifuge the contents at 2000–3000 rpm for 10 min. Remove the supernatant and wash pellets twice with buffer A (10 ml). Isolate ghosts from the pellet for extraction of lipids as described in Section 1.2.3.

Labelling of cells with TNBS

Note. Incubation of RBC with TNBS from 30 to 120 min at 4 °C is sufficient for optimal labelling of externally oriented aminophospholipids without causing labelling of intracellular compartment.

1.2.2   Labelling of Cells with Fluorescamine

Materials
- Buffer B. 100 mM KCl, 50 mM NaCl, 1 mM CaCl$_2$, 1 mM MgCl$_2$, 5 mM NaHCO$_3$, and 20 mM Tricine, pH 8.0.
- Buffer C. 100 mM KCl, 50 mM NaCl, 1 mM CaCl$_2$, 1 mM MgCl$_2$, 6 mM glycylglycine, and 20 mM Tricine, pH 8.0.
- Fluorescamine (120 mM; Sigma Chemical Co.) dissolved in 1 part of dimethlysulphoxide (DMSO) and 2.5 parts of acetone (v/v).

Labelling of cells with fluorescamine

1. To an aliquot of RBC (3 ml) suspended in buffer B add fluorescamine (8–20 μmol/μmol of RBC phospholipid). Vortex the tube immediately for exactly 30 s and add 7.5 ml of buffer C to terminate the reaction. Each sample tube is to be separately vortexed for 30 s.

2. Centrifuge cells at 2000 rpm for 10 min. Discard the supernatant.

3. Lyse the cells by adding 350 μl of 20 mM glycylglycine. Add 3 ml of chloroform:propan-2-ol (7:11, v/v) mixture to the tube for extraction of lipids as described by Rose and Oklander (1965).

1.2.3   Lipid Extraction

Lipid extraction

1. Add to red blood cells 20 volumes of 5 mM Tris buffer, pH 8.0, containing 1 mM ethylene-diaminetetracetic acid (EDTA) and keep it on ice (4 °C) for 20 min to lyse the cells.

2. Centrifuge the lysed cells in a Corex tube at 10 000 rpm for 30 min. Discard the pink-red coloured haemoglobin containing supernatant by careful aspiration with a suction pump. Carefully dislodge the pinkish ghosts from the shiny button composed of white blood cell proteases.

   To pooled pinkish ghosts, add 10 volumes of 5 mM Tris-EDTA buffer. Centrifuge at 10 000 rpm for 30 min.

Discard supernatant and collect slightly pinkish ghosts for extraction of lipids. If ghosts are still pinkish, add Tris-EDTA buffer and centrifuge again at 10000 rpm for 30 min. Suspend ghost pellet in 0.5 ml of Tris-EDTA buffer.

3. To 0.5 ml of ghost pellet add 9 ml of chloroform:propan-2-ol (7:11, v/v) in a Corex tube. Cover the tube with the cork and shake for 2 h.

4. Add 1.4 ml of 50 mM KCl, shake contents for 2 h or, if necessary, keep it overnight. Centrifuge at 10000 rpm for 20 min. Aspirate the lower lipid layer with a fine Pasteur pipette and filter the lipid extract through Whatman #1 paper to remove extraneous protein.

   Dry the sample under an atmosphere of nitrogen. Dissolve the dried sample in 100 μl of chloroform for analysis by TLC.

1.2.4 Phospholipid Analysis

1. The individual phospholipids can be separated by thin layer chromatography (TLC) on Silica gel plates (250 μm thick, 19 channels; Si 259 baker, Philipsburg, NJ) utilising the following solvent systems for one-dimensional chromatography: chloroform:methanol:acetic acid:water (50:25:8:4, v/v; Jain 1984) or chloroform:methanol:formic acid (65:25:5, v/v; Daleke and Heustis 1985).

   For better resolution from other minor phospholipids, one can utilise two-dimensional TLC utilising the following solvent systems chloroform:methanol:acetic acid:water (50:25:8:4, v/v) in the first direction and chloroform:methanol:water (5:10:1, v/v) in the second direction (Jain 1984); or chloroform:methanol:ammonium hydroxide (25%):water (180:108:11:11, v/v) in the first direction and chloroform:methanol:acetic acid:water (90:40:12:2, v/v) in the second direction (Kuypers et al. 1984).

Phospholipid thin layer chromato-graphy

2. Dry the TLC plate in air and then place it in a chamber of iodine vapours (prepared by adding 1–2 g of solid iodine in an old chromatographic chamber and heating the chamber in an oven to approx. 60 °C for 5 min so as to create vapours of iodine). Phospholipid spots will appear as yellow; circle these spots with needle.

3. Scrap the silica gel from one spot at a time with a razor blade. Transfer the scrapped gel by tapping the upright plate on to weighing paper/mylar paper. Transfer the silica gel contents of each spot scrapped into a clean tube for ashing with perchloric acid for the estimation of inorganic phosphate (as detailed in Chap. IV, this Vol.).

1.3   Tips and Precautions

- Phosphate buffer should be avoided during treatment with TNBS as it increases permeability of TNBS. Avoid the use of albumin since it blocks the reaction of TNBS with amino groups in proteins and aminophospholipids. Do not treat cells or membranes with TNBS at 37 °C, as TNBS permeability increases and thus will label intracellularly localised aminophospholipids.
- Use freshly prepared TNBS as it undergoes 8–10% hydrolysis in bicarbonate buffer when kept for 20–22 h at room temperature.
- Since most fluorescent labels are light-sensitive, it is preferable to perform labelling with fluorescamine in dark. Avoid addition of excess probe and acidic conditions, as these could lead to an artefactual labelling. Since fluorescamine reacts very rapidly, i.e. in less than 1 s with amino groups, for labelling of externally oriented aminophospholipid the time of 30 s is more than enough. If labelling is carried out for a longer time, the probe can react with intracellular lipids. If labelling is carried out at 4 °C there is a reduced chance of labelling of intracellular lipids.

2 Enzymatic Probes to Study Lipid Asymmetry

G.M. GUPTA

2.1 Background

While asymmetry of biological membranes is absolute for membrane bound proteins and carbohydrates (Rothman and Lenard 1977), virtually every type of phospholipid is present in both the monolayers, albeit in unequal amounts (Op den Kamp 1979). For example, in mammalian erythrocytes, choline-containing phospholipids are localised mainly in the outer monolayer, whereas aminophospholipids are present almost exclusively in the inner monolayer. This phospholipid asymmetry in these cells seems essential not only for maintaining the normal structure and function of the erythrocyte membrane, but also to circumvent both the red cell destruction by the spleen (Schroit et al. 1985) and hyperactivation of the blood coagulation system (Zwaal et al. 1977).

Earlier studies have revealed that the membrane phospholipid asymmetry in native erythrocytes could have been maintained primarily by the differential interactions of membrane phospholipids with membrane skeletal proteins (Haest 1982; Gupta 1987; Williamson et al. 1987). Alternatively, an ATP-dependent aminophospholipid pump which translocates phosphatidylethanolamines and phosphatidyl-serines from the outer to the inner monolayer has been considered as the sole factor in determining the asymmetric transbilayer phospholipid distribution in erythrocytes (Seigneuret and Devaux 1984; Devaux 1991). However, there is now increasing evidence to suggest that both the membrane bilayer-skeletal protein interactions and ATP-dependent aminophospholipid pump are required for maintaining the transbilayer phospholipid distribution in erythrocytes (Gupta 1992). While the aminophospholipid appears to be essential for

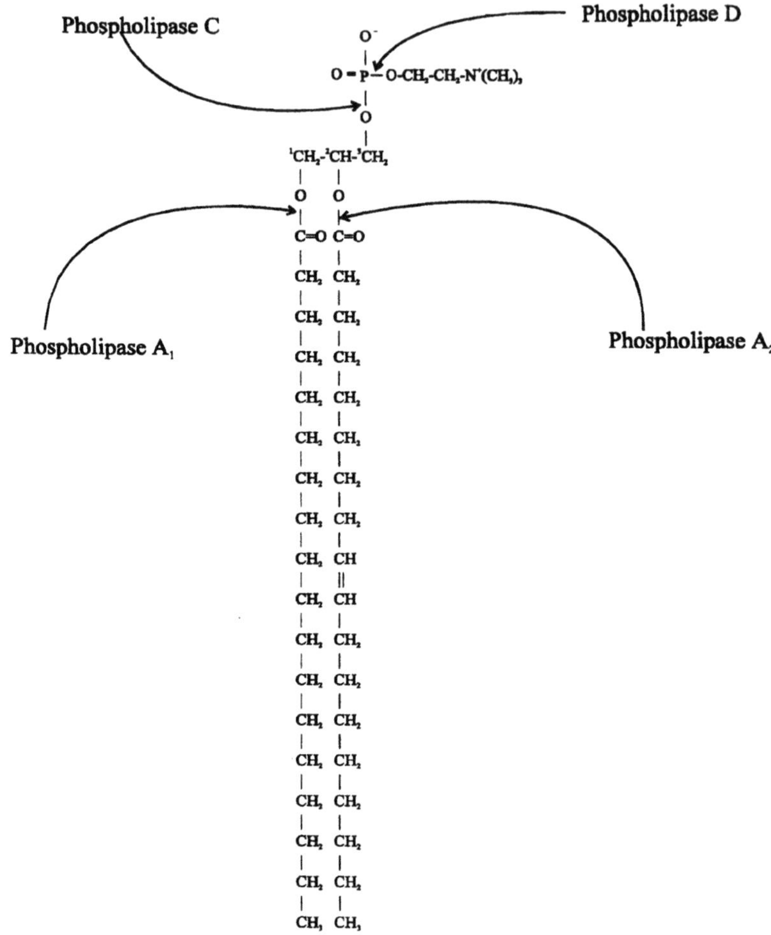

Fig. 1. Sites of action of different phospholipases on phosphatidylcholine

generation and, later, for maintaining the asymmetry, the membrane skeleton-inner layer phospholipid interactions are, perhaps, important in stabilising the aminophospholipid arrangement in the inner monolayer (Kumar et al. 1990; Loh and Huestis 1993; Muller et al. 1993).

A variety of techniques have been employed for probing the transbilayer phospholipid asymmetry in biological membranes; but amongst these, only phospholipases and amino-group labelling reagents (discussed in the preceding Section A) have found frequent use in studies of transmembrane phospholipid distribution. Figure 1 depicts the sites of action of various phospholipases on a glycerophospholipid. While phospholipase A_2 from different sources has been utilised to analyse the glycero-phospholipid distributions, sphingomyelinase C from Staphylococcus aureus has been used to determine the sphingomyelin distribution across the membrane.

Phospholipase A_2, specifically cleaves the C-2 ester bond in diacylglycerophospholipids, e.g. phosphatidylcholines, phosphatidylethanolamines, phosphatidylserines, phosphatidyl inositols etc., resulting in the formation of lysophospholipids. Since this enzyme does not permeate the membrane under normal conditions due to its macromolecular nature, its action is usually confined to the outer surface phospholipids in intact cells. The amount of external glycerophospholipids could thus be determined by quantitating the amounts of lysophospholipids formed after the enzyme treatment.

Phospholipases A_2 from different sources have different substrate specificties. (Roelofsen 1982). Thus, Naja naja phospholipase A_2 hydrolyses phosphatidylethanolamines faster than phosphatidylcholines and phosphatidylserines. Likewise, phosphatidylcholines and phosphatidylethanolamines are hydrolysed faster than phosphatidylserines by bee venom phospholipase A_2, whereas phosphatidylserines are degraded faster than phosphatidylcholines and phosphatidylethanolamines by phospholipase A_2 from porcine pancreas.

Phospholipase A_2 from porcine pancreas does not hydrolyse fresh normal erythrocytes, but it readily degrades modified red cells (Haest et al. 1978; Kumar and Gupta 1983; Chandra et al. 1987; Joshi et al. 1987; Kumar et al. 1990). The degree to which phospholipases A_2 hydrolyse various glycero-phospholipids of intact normal cells can usually be correlated with the extent of their localisation in the outer monolayer (Op den Kamp 1979; Roelofsen 1982). However, this correlation might not hold valid in the case of modified cells (Op den Kamp et al. 1985). Thus, the human erythrocytes, after diamide treatment, were reported to have, based on phospholipase A_2 digestion, an altered phospholipid distribution (Haest et al. 1978), but later studies using chemical and other probes showed no change in the transbilayer phospholipid distribution in diamide-treated red cells (Franck et al. 1986; Middelkoop et al. 1989).

Similarly, problems were encountered with the use of phospholipase A_2 in the case of sickled erythrocytes (Op den Kamp et al. 1985). Further, conflicting reports appeared on transbilayer phospholipid distributions in malaria-infected erythrocytes (Gupta and Mishra 1981; Joshi et al. 1987; Van der Schaft et al. 1987; Joshi and Gupta 1988; Moll et al. 1990), but recent studies of Sherman and coworkers (Maguire et al. 1991) have validated the earlier reports of Gupta and coworkers (Gupta and Mishra, 1981; Joshi et al. 1987; Joshi and Gupta 1988). Keeping these problems in view, it is strongly suggested that an alternate probe, e.g. amino-group labelling reagents, fluorescent probes, phospholipid exchange protein etc., should invariably be used whenever phospholipase A_2 is employed as an external membrane probe to confirm the membrane phospholipid organisation in modified erythrocytes.

Digestion of intact cells with phospholipase A_2 alone does not ensure complete hydrolysis of all external glycerophospholipids. This could, however, be achieved by first treating the cells with sphingomyelinase C, which hydrolyses the external sphingomyelins, leading to an impairment of the membrane phospholipid packing and, consequently, an increased glycerophospholipid accessibility to phospholipase A_2 in the partially digested cells (Roelofsen 1982). Besides, phospholipase A_2 from different sources may be used to overcome the problem of their preferential behaviour towards various glycerophospholipids. Also the phospholipases A_2, having different substrate specificities, may be combined together to reduce the effective digestion time.

2.2 Materials and Methods

2.2.1 Digestion of Cells with Phospholipase A_2

* Phospholipases A_2 from porcine pancreas or bee venom (Sigma Chemical Co.)
* Buffer A. 10 mM glycylglycine containing 100 mM KCl, 50 mM NaCl, 0.25 mM $MgCl_2$, 0.25 mM $CaCl_2$ and 44 mM sucrose; pH 7.4
* Buffer B. 10 mM glycylglycine containing 100 mM KCl, 50 mM NaCl, 0.25 mM $MgCl_2$, 10 mM $CaCl_2$, 44 mM sucrose; pH 7.4

Materials

1. Centrifuge blood at 900g for 15 min. Remove the plasma and buffy coat from the top of the erythrocytes by a Pasteur pipette. To the pellet add 5 ml of isotonic saline (145 mM NaCl) and centrifuge at 900g for 15 min. Discard the supernatant and repeat this procedure for two more times.

Digestion with Phospholipase A_2

2. To completely remove the leukocytes, pass the above erythrocyte preparation, after suspending in 2 ml of isotonic saline, through a 2-ml cellulose CF-11 column. Pellet the cells by centrifuging the cell suspension, derived after passing through the CF-11 column, at 900g for 30 min, and discard the supernatant.

3. Suspend an aliquot of 0.25 ml of packed erythrocytes in 5 ml of buffer A. Treat this sample as the control.

4. Add 5 to 20 IU of bee venom phospholipase A_2 to a suspension of 0.25 ml of packed erythrocytes in 5 ml of buffer A.

Note. Usually, lower amounts of phospholipase A_2 are required to hydrolyse the modified erythrocytes, compared to normal red blood cells.   **!**

5. Incubate the mixture at 37 °C for 10 min to 1 h.

! **Note.** Usually, shorter incubation times are required to hydrolyse the modified red cells, compared to normal erythrocytes.

6. Centrifuge the mixture and store the supernatant in a clean glass tube.

7. Add to the cell pellet, isotonic saline (5 ml) containing 5 mM EDTA or 2 mM o-phenanthroline and centrifuge at 900 g for 15 min. Repeat this procedure for one more time. Extract lipids directly (without lysis) from the phospholipase-treated cells by the method of Rose and Oklander (1965).

8. Measure A_{418} of red colour in supernatant of step 6 to determine the degree of haemolysis. Supernatant derived from 100% haemolysed cells is used as the control. The complete haemolysis is achieved by treating 0.25 ml of packed erythrocytes with 5 ml of distilled water.

9. For treatment with pancreatic phospholipase A_2, replace buffer A with buffer B. The remaining steps are identical to those described above for bee venom phospholipase A_2.

2.2.2   Digestion of Cells with Phospholipase A_2 and Sphingomyelinase C

Materials • Phospholipase A_2 from bee venom (Sigma Chemical Co.). Sphingomyelinase C purified from S. aureus according to Zwaal et al. (1975).
• Buffer A. 10 mM glycylglycine containing 100 mM KCl, 50 mM NaCl, 0.25 mM $MgCl_2$, 0.25 mM $CaCl_2$ and 44 mM sucrose; pH 7.4.

Digestion with phospholipase A_2 and sphingomyelinase C

1. Suspend an aliquot of 0.25 ml of packed erythrocytes in 5 ml of buffer A and treat this sample as the control.

2. Suspend another aliquot of 0.25 ml of packed erythrocytes in 5 ml of buffer A and add 5 to 20 IU of bee venom phospholipase A_2 and 10 IU of sphingomyelinase C.

Note. The use of sphingomyelinase C with bee venom   !
phospholipase A_2 leads to an extensive haemolysis in the case
of chemically treated or pathological samples of RBCs, like
malaria-infected erythrocytes, erythrocytes of patients with
chronic myeloid leukaemia and Ca^{2+}-loaded human erythrocytes, etc.

3. Incubate the mixture at 37 °C for 10 min to 1 h.

4. Centrifuge the mixture and store the supernatant in a clean
    glass tube.

5. Stop the enzyme reaction and extract lipids from washed
    cells as described above in Section 2.2.1, Digestion of Cells
    with Phospholipase A_2 (Step 7).

6. Determine the extent of haemolysis in supernatant of Step 4
    as described above in Section 2.2.1, Digestion of Cells with
    Phospholipase A_2 (Step 8).

2.2.3 Lipid Extraction

- Freshly distilled propan-2-ol   **Materials**
- Methanol
- Chloroform
- Magnetic stirrer.

All freshly distilled and of analytical grade.

1. Add 5 ml of propan-2-ol to sodium chloride (0.9%) washed   **Lipid**
    enzyme-treated or untreated erythrocytes, and stir the mix- **extraction**
    ture on a magnetic stirrer for 1 h.

2. To the above mixture, add 2.5 ml of chloroform and con-
    tinue the stirring for another 1 h.

3. Filter the mixture through a coarse sintered funnel under
    mild vacuum.

4. Remove the solvents from the filtrate below 40 °C under vacuum.

5. Dry the residue under vacuum overnight in a vacuum desiccator.

6. Dissolve the dried lipid mixture in 0.2 to 0.5 ml of freshly distilled and dry chloroform.

7. Pass the solution through an Eppendorff tip tightly plugged with clean cotton.

8. Remove the solvent and redissolve the residue in 0.1 to 0.2 ml of chloroform:methanol (1:1, v/v). This solution is directly used for phospholipid analysis by thin-layer chromatography (TLC).

2.2.4 Phospholipid Analysis

Materials
- Silica gel 60F–254 (0.25 mm thick)
- Glass or plastic-backed plates (E. Merck)
- Distilled and dry chloroform and methanol.

TLC of phospholipids
1. The individual phospholipids can be separated by multiple-dimensional TLC according to Pollet et al. (1978). After applying the phospholipid to the bottom left corner of a 10-cm × 10 cm TLC plate in the form of a streak over 1 cm, develop the plate as follows:
   (a) in chloroform:methanol:water (70:30:4, v/v) from bottom to the top,
   (b) rotate the plate counter clockwise by 90° and then develop up to two thirds of the plate length in the second solvent system, chloroform:methanol (2:8, v/v),
   (c) in the same direction from bottom to the top in chloroform:methanol (2:1, v/v), And finally
   (d) rotate the plate clockwise to return to the original position and then develop it from bottom to the top in chloroform:methanol (2:1, v/v). Dry the plate at about 40 °C in between each development.

2. Dry the TLC plate in air and then place it in an iodine chamber. The phospholipid spots will appear as yellow. Circle these spots with needle.

3. Scrap the silica gel from one spot at a time with a razor blade. Transfer the scrapped silica gel by tapping the upright plate onto a weighing paper. Transfer the silica gel contents of each spot to clean glass tubes for digestion with perchloric acid and estimation of inorganic phosphate (see Chap. IV, this Vol.).

   The percentage of phospholipid hydrolysed after phospholipase A_2 treatment is determined by measuring the ratio of remaining diacylglycerophospholipid to the corresponding lyso-derivative. The sphingomyelin degradation by sphingomyelinase C is determined by comparing the absolute and relative quantities of sphingomyelins recovered from the enzyme-treated sample with the absolute and relative quantities of sphingomyelins recovered from the untreated control sample.

2.3 Tips and Precautions

- Extraction of phospholipids from the phospholipase-treated cells should invariably be carried out from the intact cells, rather than from their ghosts, as additional phospholipid hydrolysis may occur during the ghost preparation.
- In some cases, use of EDTA alone may not be sufficient to inhibit the enzyme activity. It is, therefore, advisable in such cases to use both EDTA and o-phenanthroline to inhibit the enzyme.
- The enzyme quantity, Ca^{2+} concentration and incubation time may be varied to find out the optimum conditions where maximum degradation of external phospholipids is affected without lysing the cells to a significant extent ($\leqslant 5\%$).

3   Use of Paramagnetic and Fluorescent Analogues to Study Transmembrane Distribution and Movement of Phospholipids

A. ZACHOWSKI AND T. POMORSKI

3.1   Background

The distribution of lipids, and more especially phospholipids, between the two leaflets of biological membranes has been studied in a variety of cells over the past 20 years, since the first indication of an asymmetrical arrangement of phosphatidyl-ethanolamine (PE) in the human erythrocyte membrane (Bretscher 1972). As discussed in preceding sections, most of these determinations were based on modifications of the phospholipids present in the outer leaflet either by chemical reagents or by action of phospholipases. As a matter of fact, only the steady-state distribution could be revealed from these studies, and no indications on the rate of transmembrane movement could be available. However, in some instances, the phospholipid distribution was inferred from experiments us-ing a phospholipid exchange protein. From the kinetics of ex-change of phospholipids, it was sometimes possible to measure the inner to outer leaflet rate of movement of the phospholipid under study (Zachowski 1993). The indication that phospholipids diffuse through the bilayer and that they can adopt an asymmetric topology raises the question of the mechanism(s) maintaining this distribution. Schematically, either there exists some specific interactions between a given phospholipid and protein [as proposed for phosphatidylserine (PS) and spectrin in the red cell] or the distribution is in a dynamic equilibrium between the transverse movements of all the phospholipids. The answer to this old debate was obtained from studies of the behaviour of the four main phospholipids (PC, PE, PS and sphingomyelin, SM) as revealed by analogues which report on the dynamics of the membrane molecules.

3.2   Objectives and Limitations

Analogues to be used in the kinetic studies are subjected to several requirements:

- They cannot be modified at their polar head to ensure the specificity of this moiety.
- They have to incorporate readily into the outer membrane leaflet in order to get a zero time of the kinetics.
- One should be able to determine quickly the amount of analogue in both membrane monolayers at any time.

In fact, two related analogue families are used for this purpose, either paramagnetic with a doxyl ring or fluorescent with a 7-nitrobenz-2-oxa-1,3-diazol-4-yl (NBD) moiety. The reporter group appears on a short chain (pentanoic, caproic or dodecanoic acid) esterified at the sn-2 position of the glycerophospholipids or linked by an amide bond to the ceramide backbone. The synthesis of the paramagnetic probes has been described elsewhere (Fellmann et al. 1994). Synthesis of the fluorescent analogues also followed this protocol.

One could object that the data obtained with the analogues do not faithfully represent the behaviour of the endogenous molecules, or the behaviour of phospholipids bearing two long chains. In fact, this is not the case, as experiments with paramagnetic probes (Morrot et al. 1989), with some of the fluorescent molecules (Colleau et al. 1991; Connor et al. 1992) and with long-chain radioactive phospholipids (Tilley et al. 1986) essentially gave the similar results in the erythrocyte membrane. However, the short-chain analogues probably mimic peroxidised phospholipids and hence behave as substrates of the deacylase-reacylase-repairing system present in the cells. Hence, the labelled short chain can be hydrolysed (Seigneuret et al. 1984; Colleau et al. 1991). Always take into account the signal arising from the free "fatty acid" when interpreting the data.

Another important point is that the principle of the assay is the measure of the amount of analogue present in the membrane outer leaflet either by selective destruction or selective extraction (as described in Sect. 3.3). If the analogue can disappear from the outer leaflet by diffusing into the inner leaflet, there is also the possibility, especially when studying the plasma membrane of nucleated cells, that it is removed by the membrane endocytic pathway. Before drawing conclusions from the experimental data, it is extremely important either to inhibit this pathway by lowering the temperature (Martin and Pagano 1987) or to quantify it by an independent measurement (Cribier et al. 1993).

3.3   Materials and Methods

The PC analogue is synthesised from a lyso-PC molecule which is reacylated by a probe-bearing fatty acid. PE and PS analogues are obtained after exchange of the choline head group by an ethanolamine or serine catalysed by phospholipase D (Fellmann et al. 1994). The sn-2 chain is either a (4-doxyl)-pentanoic acid in the case of paramagnetic analogues or a (6-NBD)-caproic or a (12-NBD) dodecanoic acid in the case of fluorescent analogues. The corresponding sphingomyelin (SM) analogues are obtained by attaching the labelled fatty acids to a sphingosylphosphocholine moiety.

Due to the presence of the short chain bearing a relatively hydrophilic substituent, the phospholipid analogues are not as hydrophobic as are genuine phospholipids, and form micelles when resuspended in aqueous buffers (Tanaka and Schroit 1983; Seigneuret et al. 1984). Upon addition of membranes, they will incorporate into the exposed bilayer to an extent governed by their partition coefficient and the relative membrane volume. One has to adjust this latter parameter in order to obtain a maximum incorporation.

The amount of analogue still present in the outer membrane layer at any time of the reorientation kinetics is estimated either by its selective destruction in this layer (reduction of the paramagnetic molecule by ascorbate or chemical modification by dithionite of the fluorescent group into a non-fluorescent one) or by its extraction (back-exchange) from this layer by hydrophobic acceptors such as serum albumin or liposomes. Here we will describe only the use of albumin in such experiments. For back-exchange on liposomes, refer to Connor et al. (1992).

The method is revised from Fellmann et al. (1994). It is described here for red blood cells and nucleated cells. When using organelles, conditions have to be chosen to work with the same phospholipid concentration. Accordingly, the centrifugation conditions to pellet the membranes have to be modified and sometimes it may be advantageous to use a very high-speed centrifuge (Beckman Ti-100, for instance) instead of a bench-top microcentrifuge.

Cells

Cells or organelles are suspended in the desired buffer at a concentration which depends on the material available. For instance, when using erythrocytes, it is convenient to resuspend them in a 50% hematocrit. It is obvious that such a cell concentration might be difficult to reach in a volume sufficient for the experiments when cultured cells are used.

When the experiment is to be carried out with spin-labelled lipids, we recommend a minimal cell concentration of [2 to 3] $\times$ 10^8 cells per ml. When fluorescent labels are used, the cell concentration may be lowered due to the higher sensitivity of fluorescence over ESR.

The suspension medium should not contain serum albumin or calf serum, as these proteins bind to the phospholipid analogues.

When the hydrolysis of the analogues by phospholipase A_2-like enzymes present in the biological sample is important, cells or organelles may be treated by 5 mM (final concentration) diisopropyl-fluorophosphate (DFP) for 10 min at 37 °C in order to (partially) inhibit the hydrolytic activity. However, one should verify that this treatment does not change the cell integrity or impair its functions.

Markers   The desired amount of phospholipid analogue in organic solution (corresponding to 1 to 3% of the endogenous phospholipids in the case of the paramagnetic analogues and at least to 0.1% of these in the case of the fluorescent analogues) is deposited in a glass tube, dried under reduced pressure and resuspended in the experimental buffer by vigorous vortexing. The volume has to be chosen according to the experiment (see below).

With the fluorescent analogues, addition of a small volume of ethanol to the dry film and a brief sonication under nitrogen after adding the buffer can help to resuspend all the probes.

Reagents
- 300 mM sodium ascorbate solution is prepared in the desired buffer when experiments are run with paramagnetic probes. Take care to check, and adjust if necessary, the pH of this solution.
- 1 M sodium dithionite solution is prepared in 1 M Tris (pH 10) when using fluorescent molecules (McIntyre and Sleight 1991). This solution has to be kept on ice and used within 3 h.
- Fatty acid free bovine serum albumin is dissolved at 5% (w/v) in the experimental buffer.
- 100 mM potassium hexaferricyanide solution.

Equipment
- 5-ml disposable glass tubes and 1.5-ml microcentrifuge tubes.
- Micropipettes with tips.
- Ice bath and thermostat water bath.
- Bench-top microcentrifuge.
- Electron spin resonance (ESR) or fluorescence spectrometer.
- Vortex mixer.

Cell labelling   Mix in a glass tube, cell or membrane suspension (2 Vol.) and marker suspension (1 Vol). Heat all the suspensions to the experimental temperature before mixing. This step is the time zero of the reorientation kinetics.

With erythrocytes, 90 µl aliquots are sampled from the incubation at given times and added to a microcentrifuge tube containing 20 µl of serum albumin solution usually chilled to 4 °C on an ice bath. Shake the tube to ensure mixing and leave for 1 min on ice.

Back-exchange on serum albumin

Note. This incubation is sufficient to allow the extraction of the outwardly exposed analogue and is too short to allow a significant movement of the analogue from the inner to the outer membrane leaflet at this low temperature.

!

Centrifuge the suspension to pellet the cells. An aliquot of the supernatant is pipetted out and frozen. Alternatively, remove all the supernatant and save pellet.

Note. The estimation of the analogue present in the fraction depends on the probe used. In the case of a less concentrated cell suspension or of membrane suspension, the volume of the aliquot may be reduced to 50 µl.

!

Paramagnetic analogue. The amount of spin-labelled analogue present in the albumin-containing supernatant or in the cell pellet is determined by recording their ESR spectrum and comparing them either to the spectrum from the extract at time zero or to a pellet obtained without addition of albumin, respectively. It generally happens that probes are reduced with time by the cytoplasmic content and thus the recorded signal is artificially diminished. One way to bypass this problem is to reoxidise the probes prior to spectroscopic measurement by adding ferricyanide (10 mM final concentration) to the sample to be measured.

Fluorescent analogue. One can choose to measure the fluorescence emitted either by the supernatant containing the probes extracted from the outer leaflet or by the cell pellet with the probes located in the cytoplasmic leaflet. In the latter case, it may be worth solubilising the membranes by adding detergent to the suspension. This leads to obtaining a higher signal by dilution of the probes in micelles, thus diminishing self-quenching, which could occur in the bilayer by dispersion of any quenching molecule such as haemoglobin and by a higher quantum yield (Tanaka and Schroit 1983).

Addition of detergent to the supernatant may also be convenient if one wants to compare it to the solubilised pellet. The fluorimeter settings are: excitation wavelength at 470 nm and emission wavelength at 540 nm.

Reduction of paramagnetic analogues by ascorbate This method can be used solely when the analogues are not reduced by the cell content over the incubation time (e.g. erythrocytes incubated at 4 °C). Nitroxide radical is reduced to a non-paramagnetic species by ascorbate (Kornberg and McConnell 1971). This property has been used to quantify the amount of spin labelled analogue present in the outer membrane layer and thus accessible to ascorbate.

After addition of the reducer to the membrane suspension, the signal intensity decreases then plateaus when all the outwardly oriented paramagnetic molecules are reduced. The remaining signal arises from the molecules located in the inner leaflet. However, cells such as erythrocytes are relatively permeant to ascorbate and even the labels positioned in the inner layer can be destroyed. Hence, measurements are made at low temperature (i.e. 4 °C) where the ascorbate uptake is drastically reduced and only the outer layer is readily accessible to the reducer.

An aliquot of the cell suspension is transferred to an ESR cell after addition of 30 to 50 mM ascorbate and signal intensity recorded at short intervals (1 min for instance). The kinetics of signal, which decreases with time, is biphasic and the signal intensity is plotted on a semi-log scale. One can observe a fast signal decrease followed by a slow loss of signal intensity. Extrapolation of this second kinetics to time zero of the reduction process gives an estimate of the amount of probe which is localised into the inner membrane layer.

The NBD moiety loses its fluorescent property after modification by dithionite (McIntyre and Sleight 1991). The distribution of the fluorescent analogues is measured directly in the cell suspension by comparing the fluorescence intensity before and after addition of dithionite. Practically, an aliquot of the cell suspension is transferred to a fluorimeter cuvette, diluted if desired to the appropriate volume, and the signal intensity recorded. Then 100 mM dithionite is added from the stock solution and the signal decrease followed with time. The accessible fluorophores of the outer leaflet are rapidly transformed in non-fluorescent molecules and the remaining stable signal corresponds to the fraction of the analogues present in the inner leaflet.

Loss of fluorescence after addition of dithionite

Again, the membrane may be permeable to dithionite and the latter signal can also be affected by the molecules which entered into the cell through the anion transporter. One way to prevent or to reduce this undesirable effect is to perform the measurement at a low temperature (i.e. 15 °C), to use a lower dithionite concentration or to treat the cells by an anion transport inhibitor prior to the experiment. In the case of erythrocytes, the uptake of dithionite can be sufficiently impaired by incubation of the cells with 50 μM 4,4′-diisothiocyanatostilbene-2,2′-disulfonic acid (DIDS; Pomorski et al. 1994).

To ensure that the remaining signal is really due to the molecules in the inner membrane leaflet, 1% Triton X-100 can be added at the end of the measurement to solubilise membranes and give access to all the fluorophores. No more signal should be detectable after such a treatment.

It is possible to obtain a faster signal loss from paramagnetic or fluorescent molecules, after addition of ascorbate or dithionite, respectively, if the cell suspension aliquot is mixed with serum albumin solution (1 to 2% final concentration) prior to the addition of the chemicals. Probes originally present in the membrane outer leaflet and adsorbed onto albumin are much more accessible to ascorbate or dithionite, and their associated

Alternative signal reduction procedure

signal is destroyed within seconds, allowing an easier discrimination from the reduction of the signal arising from analogues situated in the inner leaflet when uptake of ascorbate or dithionite cannot be totally inhibited. The remaining signal has to be compared with the signal arising from an aliquot of the cell suspension supplemented with buffer instead of albumin and chemical reagent.

3.4 Tips and Precautions

• Control of Analogue Incorporation in Membrane

It may happen that the analogue added to the membrane suspension is not taken immediately by the outer monolayer and requires some time to reach a complete incorporation. Indeed, this will prevent one from determining the real beginning of the reorientation kinetics. Estimation of the amount of analogue still present in the external medium can be made by centrifuging an aliquot of the membrane suspension and recording the signal arising from the supernatant after addition of detergent to disperse the analogue micelles. In this case, membrane labelling is better performed at low temperature, when all the transmembrane motions are slow; the suspension is centrifuged, the supernatant discarded and the pellet resuspended in incubation buffer; the sample is then brought to the desired temperature to initiate the phospholipid transmembrane reorientation.

It is also possible to verify the complete incorporation of the paramagnetic analogues by recording the ESR spectrum of the membrane suspension: the absence of unincorporated analogue corresponds to the absence of narrow spectral lines due to monomers in solution (Seigneuret et al. 1984).

• Loss of Signal

The intensity of the signal associated with the membrane can be modified for several reasons. It may be due to a reversible reduction of the paramagnetic probe, as reported previously; then, signal can be recovered by addition of a gentle oxidant such as ferricyanide. It is obvious that under these conditions the ascorbate method cannot be applied and that only the back-exchange method is suitable.

In other cases, the loss may be irreversible as it occurs when the paramagnetic analogues are in contact with an acidic medium. Care should be taken in interpreting the data, as the destroyed analogues could be considered as localised in the membrane inner leaflet.

• Back-Exchange Efficiency

The amount of serum albumin required to totally extract the analogue from the outer membrane leaflet, as well as the length of contact required, may vary from one membrane to another. To determine the optimal conditions, it is possible to label the membrane at low temperature and to measure the amount recovered on various amounts of serum albumin and the amount left in membrane after different times of contact of the membrane suspension with serum albumin.

It may happen that at 4 °C not all the analogues are extracted from the outer membrane leaflet and one has to repeat the assay at higher temperature.

• Metabolism of the Analogue

Some membranes are equipped with enzymes which modify phospholipids, an example is the mitochondria in which PS is decarboxylated to PE. One should take care that the analogues are not modified during the course of the study using, for instance, lipid extraction in organic solvents followed by their separation by thin layer chromatography. However, such determination may be hampered by the paucity of available material due to the amount of membrane and hence, of analogue present in the incubation.

A more frequent modification is the transformation of the glycerophospholipid analogues in lyso-derivatives by the action of phospholipase A$_2$-like enzymes. The labelled short-chain fatty acids liberated are practically not soluble in a bilayer and thus appear in the aqueous medium. When the back-exchange technique is used, they will be recovered in the supernatant, even if not bound to the albumin molecule, and contribute to the recorded signal. In the case of a paramagnetic molecule, their amount can be directly estimated from the supernatant spectrum as the soluble, fast-tumbling short fatty acids will generate a narrower signal than the albumin-bound phospholipids; the two spectral components can be differentiated and quantified (Fellmann et al. 1994).

Another approach, which can be used with fluorescent analogues as well, is to take two aliquots at any time point, one being supplemented with albumin and the other with buffer. The signal arising from the second supernatant, subsequently supplemented with serum albumin, will be an indication of the amount of free fatty acid present and will serve to correct the signal given by the first supernatant.

• Accuracy of Time Zero

When applying the back-exchange protocol, it is possible that the transmembrane movement is so fast at high temperature that a fraction of the analogues is already located in the inner leaflet and cannot be extracted at the time of the first aliquot sampling. It is thus impossible to estimate the true initial velocity of the reorientation kinetics. It is recommended in this case to repeat the measurement at a lower temperature when the movement is slower.

• Cellular Energy Status

The movement and distribution of the amino-phospholipids, PE and PS, in the plasma membrane of animal cells are dependent on the hydrolysis of cytoplasmic ATP. Results obtained are thus affected by the amount of ATP available. However, during prolonged incubations, it may happen that the buffer does not contain enough energy sources (e.g. glucose) to allow the cells to maintain a constant ATP level. The nucleotide concentration has to be measured at the end of the incubation and the buffer adjusted to ensure that ATP will not decrease with time.

• Cell Integrity

Cell viability or organelle lysis have to be checked at the end of the experiment, as one cannot distinguish anymore, an inner leaflet and an outer leaflet in an unsealed membrane.

Note. Phospholipid analogues are stored at $-20\,^{\circ}\mathrm{C}$ in chloroform:methanol (1:1, v/v). Care should be taken when using DFP, as this chemical is highly toxic.

References

Blumefield N, Zachowski A, Galacteros F, Beuzard Y, Devaux PF (1991) Transmembrane mobility of phospholipids in sickle erythrocytes: effect of deoxygenation on diffusion and asymmetry. Blood 77:849–854

Bretscher MS (1972) Asymmetrical lipid bilayer structure. Nature (New Biol) 236:11–12

Chandra R, Joshi P, Bajpai VK, Gupta CM (1987) Membrane phospholipid organization in calcium-loaded human erythrocytes. Biochim Biophys Acta 902:253–262

Colleau M, Hérve P, Fellmann P, Devaux PF (1991) Transmembrane diffusion of fluorescent phospholipids in human erythrocytes. Chem Phys Lipids 57:29–37

Connor J, Pak CH, Zwaal RFA, Schroit AJ (1992) Bidirectional transbilayer movement of phospholipid analogues in human red blood cells. Evidence for an ATP-dependent and protein-dependent process. J Biol Chem 267:19412–19417

Cribier S, Sainte-Marie J, Devaux PF (1993) Quantitative comparison between aminophospholipid translocase activity in human erythrocytes and in K562 cells. Biochim Biophys Acta 1148:85–90

Daleke DL, Heustis WH (1985) Incorporation and translocation of aminophospholipids in human erythrocytes. Biochemistry 24:5406–5416

Devaux PF (1991) Static and dynamic lipid asymmetry in cell membranes. Biochemistry 30:1163–1173

Fadok VA, Voelker DR, Campbell PA, Cohen JJ, Bratton DL, Henson PM (1992) Response of phosphatidylserine on the surface of apoptoic lymphocytes triggers specific recognition and removal by macrophages. J Immunol 148:2207–2216

Farooqui SM, Wali RK, Baker RF, Kalra VK (1987) Effect of cell shape, membrane deformability and phospholipid organization on phosphate calcium induced fusion of erythrocytes. Biochim Biophys Acta 904:239–250

Fellmann P, Zachowski A, Devaux PF (1994) Synthesis and use of spin-labelled lipids for studies of the transmembrane movement of phospholipids. In: Graham JM, Higgins JA (eds) Methods in Molecular Biology, Vol. 27. Humana Press, Ottawa, pp 161–175

Franck PFH, Op den Kamp JAF, Roelofsen B, van Deenen LLM (1986) Does diamide treatment of human erythrocytes cause a loss of phospholipid asymmetry? Biochim Biophys Acta 857:127–130

Gordesky SE, Marinetti GV (1973) The asymmetric arrangement of phospholipids in human erythrocyte membranes. Biochem Biophys Res Commun 50:1027–1031

Gordesky SE, Marinetti GV, Love R (1975) The reaction of chemical probes with the erythrocyte membrane. J Membr Biol 20:111–132

Gupta CM (1987) Stabilization mechanisms of transbilayer phospholipid asymmetry in erythrocytes membrane. Curr Sci 56: 1201–1209

Gupta CM (1992) Genesis and maintenance of erythrocyte membrane phospholipid asymmetry: Comparative roles of aminophospholipid pump and membrane skeleton-bilayer interactions. In: Gaber BP, Easwaran KRK (eds) Membrane structure and functions – the state of the art. Academic Press, New York, pp 177–183

Gupta CM, Mishra GC (1981) Transbilayer phospholipid asymmetry in Plasmodium knowlesi-infected host cell membrane. Science 212:1047–1049

Haest CWM (1982) Interactions between membrane skeleton proteins and the intrinsic domain of the erythrocyte membrane. Biochim Biophys Acta 694:331–352

Haest CWM, Plasa G, Kamp D, Deuticke B (1978) Spectrin as a major stabilizer of the phospholipid asymmetry in the human erythrocyte membrane. Biochim Biophys Acta 509:21–32

Jain SK (1984) The accumulation of malonyldialdehyde, a product of a fatty acid peroxidation, can disturb aminophospholipid organization in the membrane bilayer of human erythrocytes. J Biol Chem 234:466–468

Joshi P, Gupta CM (1988) Abnormal membrane phospholipid organization in Plasmodium falciparum-infected human erythrocytes. Br J Haemat 68: 255–259

Joshi P, Dutta GP, Gupta CM (1987) An intracellular Simian malarial parasite (Plasmodium knowlesi) induces stage-dependent alterations in membrane phospholipid organization of its host erythrocyte. Biochem J 246:103–108

Kalra VK, Banerjee R, Sorgente N (1990) Heterotypic and homotypic cell-cell adhesion molecule in endothelial cells. Biotech App Biochem 12:579–585

Kornberg RD, McConnell HM (1971) Inside-outside transitions of phospholipids in vesicle membranes. Biochemistry 10:1111–1120

Kumar A, Gupta CM (1983) Red cell membrane abnormalities in chronic myeloid leukemia. Nature 303:632–633

Kumar A, Gudi SRP, Gokhale SM, Bhakuni V, Gupta CM (1990) Heat-induced alterations in monkey erythrocyte membrane phospholipid organization and skeletal protein structure and interactions. Biochim Biophys Acta 1030:269–278

Kuypers FA, Roelofsen B, Op den Kamp JAF, van Deenen LLM (1984) The membrane of intact human erythrocytes tolerates only limited changes in the fatty acid composition of its phosphatidylcholine. Biochim Biophys Acta 769:337–347

Loh RK, Huestis WH (1993) Human erythrocyte membrane lipid asymmetry: transbilayer distribution of rapidly diffusing phosphatidylserine. Biochemistry 32:11722–11726

Maguire PA, Prudhomme J, Sherman IW (1991) Alterations in erythrocyte membrane phospholipid organization due to the intracellular growth of the human malaria parasite, Plasmodium falciparum. Parasitology 102:179–186

Marinetti GV, Crain RC (1978) Topology of aminophospholipids in the red cell membrane. J Supramol Struct 8:191–213

Martin OC, Pagano RE (1987) Transbilayer movement of fluorescent analogues of phosphatidylserine, phosphatidylethanolamine at the plasma membrane of cultured cells. Evidence for a protein-mediated and ATP-dependent process(es). J Biol Chem 262:5890–5898

McEvoy L, Williamson P, Schlegel RA (1986) Membrane phospholipid asymmetry as a determinant of erythrocyte recognition by macrophages. Proc Natl Acad Sci USA 83:3311–3315

McIntyre JC, Sleight RG (1991) Fluorescence assay for phospholipid membrane asymmetry. Biochemistry 30:11819–11827

Middelkoop E, Vander Hoek EE, Bevers EM, Comfurius P, Slotboom AJ, Op den Kamp JAF, Lubin BH, Zwaal RFA, Roelofsen B (1989) Involvement of ATP-dependent aminophospholipid translocation in maintaining phospholipid asymmetry in diamide-treated human erythrocytes. Biochim Biophys Acta 981:151–160

Moll G, Vial H, Bevers EM, Ancelin M, Roelofsen B, Comfurius P, Zwaal

RFA, Op den Kamp JAF, van Deenen LLM (1990) Phospholipid asymmetry in the plasma membrane of malaria-infected erythrocytes. Biochem Cell Biol 68:579–585

Morrot G, Hérve P, Zachowski A, Fellmann P, Devaux PF (1989) Aminophospholipid translocase of human erythrocyte: phospholipid substrate specificity and effect of cholesterol. Biochemistry 28:3456–3462

Muller P, Zachowski A, Beuzard Y, Devaux PF (1993) Transmembrane mobility and distribution of phospholipids in the membrane of mouse β-thalassaemic red blood cells. Biochim Biophys Acta 1151:7–12

Op den Kamp JAF (1979) Lipid asymmetry in membranes. Annu Rev Biochem 48:47–71

Op den Kamp JAF, Roelofsen B, van Deenen LLM (1985) Structural and dynamic aspects of phosphatidylcholine in the human erythrocyte membrane. Trends Biochem Sci 10:320–323

Pollet S, Ermidou S, Le Saux F, Monge M, Baumann N (1978) Microanalysis of brain lipids: multiple two-dimensional thin-layer chromatography. J Lipid Res 19:916–921

Pomorski T, Herrmann A, Zachowski A, Devaux PF, Muller P (1994) Rapid determination of the transbilayer distribution of NBD-phospholipids in erythrocyte membranes with dithionite. Mol Memb Biol 11:39–44

Rawyler A, Roelofsen B, Op den Kamp JAF (1984) The use of fluorescamine as permeant probe to localize phosphatidylethanolamine in intact Friend erythroleukemic cells. Biochim Biophys Acta 769:330–336

Roelofsen B (1982) Phospholipases as tools to study the localization of phospholipids in biological membranes. A critical review. J Toxicol (Tox Rev) 1:187–197

Rose HG, Oklander M (1965) Improved procedure for extraction of lipids from human erythrocytes. J Lipid Res 6:528–531

Rothman JE, Lenard J (1977) Membrane asymmetry—the nature of membrane asymmetry provides clues to the puzzle of how membranes are assembled. Science 195:743–753

Schlegel RA, Prendergast TW, Williamson P (1985) Membrane phospholipid asymmetry as a factor in erythrocyte-endothelial cell interactions. J Cell Physiol 123:215–218

Schroit AJ, Madsen JW, Tanaka Y (1985) In vivo recognition and clearance of red blood cells containing phosphatidylserine. J Biol Chem 260:5131–5138

Schwartz RS, Tanaka Y, Fidler IJ, Chiu D, Lubin B, Schroit AJ (1985) Increased adherence of sickled and phosphatidylserine-enriched human erythrocytes to cultured human peripheral blood monocytes. J Clin Invest 75:1965–1972

Seigneuret M, Devaux PF (1984) ATP-dependent asymmetric distribution of spin-labelled phospholipids the erythrocyte membrane: relation to shape changes. Proc Natl Acad Sci USA 81:3751–3755

Seigneuret M, Zachowski A, Herrmann A, Devaux PF (1984) Asymmetric lipid fluidity in human erythrocyte membrane: new spin label evidence. Biochemistry 23:4271–4275

Singer SJ (1974) The molecular organization of membranes. Annu Rev Biochem 43:805–833

Stoeckenius W, Engelman DM (1969) Current models for the structure of biological membranes. J Cell Biol 42:613–646

Tanaka Y, Schroit AJ (1983) Insertion of fluorescent phosphatidylserine into plasma membranes of red blood cells: recognition by autologous macrophages. J Biol Chem 258:11335–11343

Tilley L, Cribier S, Roelofsen B, Op den Kamp JAF, van Deenen LLM (1986) ATP-dependent translocation of aminophospholipids across the human erythrocyte membrane. FEBS Lett 194:21–27

Vander Schaft P, Beaumelle B, Vial H, Roelofsen B, Op den Kamp JAF, van Deenen LLM (1987) Phospholipid organization in monkey erythrocytes upon Plasmodium knowlesi infection. Biochim Biophys Acta 901:1–14

Verkleij AJ, Zwaal RFA, Roelofsen B, Comfurius P, Kastelijn D, van Deenen LLM (1973) The asymmetric distribution of phospholipids in the human red cell membrane. A combined study using phospholipases and freeze-etching electron microscopy. Biochim Biophys Acta 323:178–193

Wali RK, Jaffe S, Kumar D, Kalra VK (1988) Alterations in the organization of phospholipids in erythrocytes as factor in adherence to endothelial cells in diabetes mellitus. Diabetes 37:104–111

Williamson P, Antia R, Schlegel RA (1987) Maintenance of membrane phospholipid asymmetry. Lipid-cytoskeletal interactions or pump? FEBS Lett 219:316–320

Zachowski A (1993) Phospholipids in animal eukaryotic membranes: transverse asymmetry and movement. Biochem J 294:1–14

Zwaal RFA, Roelofsen B, Comfurius P, van Deenen LLM (1975) Organization of phospholipids in human red cell membranes as detected by the action of various purified phospholipases. Biochim Biophys Acta 406: 83–86

Zwaal RFA, Comfurius P, van Deenen LLM (1977) Membrane asymmetry and blood coagulation. Nature 268:358–360

Basil O. Ibe

1   Background

Lipids are a broad class of biomolecules which are important component of biological membranes. They serve as biosynthetic intermediates in various life processes leading to the production of bioactive phospholipids or carriers of other biologically active compounds such as fatty acids. Among the host of fatty acids linked to phospholipids, arachidonic acid (AA) has received the widest attention (Fig. 1). Therefore, release of AA and/or its metabolites can be used to assess lipid turnover in a cell, organ or tissue in normal processes or during a pathological condition.

Arachidonic acid is released from membrane phospholipids by the action of phospholipase A_2 under normal physiological conditions, during conditions of cell injury or inflammation. It is metabolised by oxygenation into scores of biologically active lipids (Hassid and Levine 1977; Henderson 1987). Alterations in AA metabolism are used to study some pathological conditions such as asthma and allergic reactions, liver and renal disorders (Huber et al. 1989) and coronary artery disease (Willerson et al. 1989). Oxygenation of AA proceeds through three principal enzymatic pathways producing a variety of compounds which are collectively termed eicosanoids (Henderson 1987; Holtzman 1991). The enzymes involved in AA metabolism are:

Department of Pediatrics, Harbor-UCLA Medical Center, UCLA School of Medicine, Torrance, CA 90509, USA

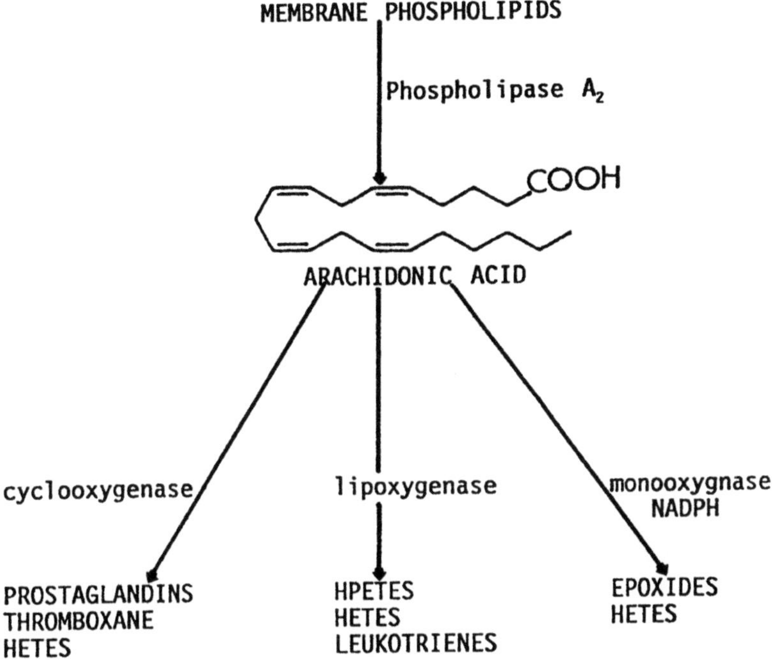

Fig. 1. Scheme of arachidonic acid metabolic pathways

- Cyclooxygenase, which produces prostaglandins (PG), thromboxane (Tx) and hydroxyeicosatetraenoic acids (HETES)
- Lipoxygenases, which produce HETES, leukotrienes (LTs) and lipoxins, and
- Monoxygenase, which produces epoxides and HETES.

One reason for the heightened interest in AA is that cellular responses to injury or inflammation, especially by the lung, can be correlated to changes in AA metabolism at the site of injury or inflammation (Henderson Jr. 1987; Holtzman 1991). The ability to measure the release of AA and its numerous metabolites during pathological conditions provides a method of understanding the role of these lipid mediators. In this chapter, we will detail the most versatile techniques used to measure AA metabolites. The discussion will concentrate on measurement of eicosanoids: PGs, Tx, HETES and LTs by RIA and HETES or LTs by HPLC. Indeed, a comprehension of the methods to be discussed will permit one to extrapolate the techniques for use in measuring any AA metabolite or lipid metabolite of interest.

2   Measurement of Eicosanoids by Radioimmunoassay (RIA)

Measurement of eicosanoids by RIA will be described with reference to PGs and Tx. PGs and Tx can be quantitated by a variety of physical and chemical methods. The physical methods which have been used include gas chromatography/mass spectrometry (Fitzpatrick et al. 1977a,b; Strife and Murphy 1984), and high performance liquid chromatography (HPLC) with either fluorescent or ultra violet (UV) detector (e.g. Turk et al. 1978). The predominant methods of chemical measurement are enzyme-linked immunosorbent assay (ELISA) and radioimmunoassay (RIA; Campbell and Ojeda 1987). The physical methods of measurement are time-consuming, require some structural modification of the PGs or Tx, are usually expensive and not quite versatile. Therefore, in terms of economy and versatility, immunoassay methods are preferred. They are highly specific, sensitive and easy to use.

2.1   Sample Preparation

The choice of a method of preparing samples for RIA measurements depends on sample media. In most cases, if the sample medium is a buffer containing no protein, RIA measurement can be done directly on the sample medium. However, if the sample medium is a biological fluid such as plasma, urine, saliva, culture media, a buffer containing protein or other lipids, an extraction should be done to remove these compounds, which may interfere with the RIA measurement. The preferred method of extraction employs a solid support as described below (Powell 1980).

- Sample media for PG or Tx measurement
- Disposable octadecylsilyl (C_{18}) column cartridge, 6-ml volume (J. T. Baker, #7020-6)
- Low pressure extraction manifold (Sep-Pak cartridge rack, Waters Associates)
- 15, 85, 95% ethanol in water
- Toluene
- Glacial acetic acid
- Ethyl acetate
- Deionised distilled water
- 13 × 100 mm glass vials pre-rinsed with methanol.

Materials

1. Add known amount of PG standard, usually 3000 cpm of [^{3}H]-PG to each unknown to determine recovery of sample from extraction process. Adjust the pH of each sample medium to 3.0 with glacial acetic acid.

2. Place the 6 ml C_{18} column, one for each sample, in the Sep-Pak cartridge rack and wash column with 5 ml of 95% ethanol.

3. Place sample on extraction column and load in column by pulling with low vacuum. Wash column sequentially with 5 ml each of 15% ethanol and 5 ml water. The column should not run dry. Effluent of steps 2 and 3 should be discarded.

4. Elute PGs and Tx with 5 ml of ethyl acetate. Collect sample eluates in the methanol rinsed 13 × 100-mm glass tubes.

5. Evaporate ethyl acetate under nitrogen at 35–40 °C. Reconstitute sample residue in 1 ml of an RIA buffer. Determine the sample recovery from extraction by counting an aliquot

Sample preparation

of the resuspension. Recovery of PGs and Tx is usually 95 ± 2%.

! **Note.** To reuse the column each column is washed with 5 ml of 85% ethanol. Each column can be reused five to ten times. However the extraction efficiency decreases with each reuse. With plasma, microsomes, cytosol or tissue homogenate samples, columns can only be reused two to three times before they become heavily clogged up. Therefore, for samples in these types of media, one column per sample is recommended.

2.2 Radioimmunoassay of PGs and Tx

To measure PGs or Tx by radioimmunoassay, a standard calibration curve is set up from which the amount of prostanoid in the sample is determined. One important point to remember is to be sure that the antisera, the non-radiolabelled PG standard and the radiolabelled metabolite are of the same PG that is to be measured. Occasional users of RIA methods do not have to grow their own antisera. PG and Tx antisera of good quality are commercially available (for instance, Caymen Chemical for antisera and ELISA kits and PerSeptive Diagnostics for antisera). These antisera come with instructions with regards to antibody titre, cross-reactivity with other eicosanoids, whether monoclonal or polyclonal and expiry date. The terms antisera and antibodies will be used interchangeably in this discussion.

Materials
- 12×75-mm polypropylene tubes
- Micropipettors (one each of 20, 100, 200, and 1000 µl capacity with matching tips)
- Eppendorff pipette
- Ice bath
- Assay buffer

- Dextran-coated charcoal
- Antisera (purchased commercially)
- Non-radiolabelled (cold) standards of PGs, Tx and HETES [³H]-PG
- Liquid scintillation cocktail
- β-counter.

1. Arrange the tubes for the standard curve and the samples **Radio** as shown in Table 1. Both the standard curve and the **immunoassay** sample tubes are set up in duplicates.

2. Set up the standard curve as described in Table 1. The volume of each concentration of standard must be 0.1 ml. The standard curve is set from 0 to 200 pg.

3. Aliquot 0.1 ml of sample solution from the extraction step in tubes, one duplicate set of tubes per sample. The range of standard concentration used have been proven to be adequate to calculate the amount of metabolite in most samples. If a sample is suspected to contain a high concentration of metabolites, it will be necessary to dilute the sample before the assay is done.

4. Dilute the antisera in the RIA buffer according to the direction provided by the manufacturer. Be sure to dilute enough antisera for each assay.

5. Dilute the [³H]-tracer of the PG to be assayed so that the radioactivity is between 8000 an 10000 cpm/0.1 ml. The organic solvent in which the tracer is suspended should be evaporated before the aliquot of tracer is taken up in the RIA buffer. Be sure to dilute enough tracer for each assay. A prior calculation of the amount needed is necessary.

6. Aliquot 0.1 ml of the antisera as described in Table 1.

Note. Aliquots of antisera should not be added in the Total and **!** Non-specific Binding (NSB) tubes.

Table 1. Procedure for radioimmunoassay standard curve and sample set up

Tube #	Buffer (ml)	Metabolite standard or sample added (ml)	Antibody (ml)	[³H]-Tracer (ml)	Explanation of amounts of metabolite added
1	0.4			0.1	Total count tubes
2	0.4			0.1	Total count tubes
3	0.4			0.1	Total count tubes
4	0.4			0.1	Total count tubes
5	0.2			0.1	Non-specific binding
6	0.2			0.1	Non-specific binding
7	0.1	0.1	0.1	0.1	0
8	0.1	0.1	0.1	0.1	0
9		0.1	0.1	0.1	5
10		0.1	0.1	0.1	5
11		0.1	0.1	0.1	10
12		0.1	0.1	0.1	10
13		0.1	0.1	0.1	25
14		0.1	0.1	0.1	25
15		0.1	0.1	0.1	50
16		0.1	0.1	0.1	50
17		0.1	0.1	0.1	75
18		0.1	0.1	0.1	75
19		0.1	0.1	0.1	100
20		0.1	0.1	0.1	100
21		0.1	0.1	0.1	200
22		0.1	0.1	0.1	200
23		0.1	0.1	0.1	Unknown samples
24		0.1	0.1	0.1	and control
25		0.1	0.1	0.1	samples or RIA
26		0.1	0.1	0.1	measurements
ETC					

Each step is peformed in duplicate. The total count tubes are in quadruplicate. Antibody is used in the dilution that binds 50% of the [³H]-tracer, its titre, or as specified by the manufacturer. Amount of metabolite used in standard curve is in picograms (pg).

7. Using the Eppendorf pipette, aliquot 0.1 ml of the [³H]-tracer into all assay tubes as indicated in Table 1.

8. Shake the tube rack gently for 30 s to ensure that components of the assay tubes are properly mixed.

9. Cover the tubes with parafilm and incubate as suggested by the manufacturer of the antisera. Usually an incubation of 8–16h at 4°C is necessary to ensure a good result of the binding assay.

10. After 8–16h incubation at 4°C, add 0.2ml of the dextran-coated charcoal to the assay as described in Table 1.

Note. Be sure that the charcoal suspension is not added to the **!** tube containing the assay for total radioactivity count.

11. Vortex the mixture of assay charcoal suspension and centrifuge the tubes at 3000rpm for 15min at 4°C.

12. After centrifugation, decant the supernatants into clean 4-ml scintillation vials (plastic or glass), add 3ml of scintillation cocktail to each assay tube and determine the radioactivity by counting on a β-counter for 5min.

Several computer programmes are available for calculating **Calculation** RIA data. Some of the programmes can perform linear as well

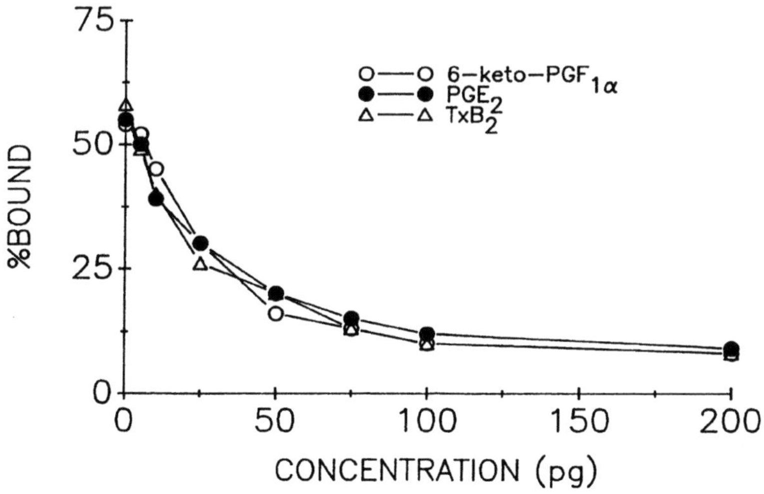

Fig. 2. Typical radioimmunoassay non-linear standard curve for 6-keto-PGF$_{1\alpha}$, TxB$_2$ and PGE$_2$. Data are means of four determinations

as logarithmic calculations. In most cases, the programme plots the percentage of radioactivity bound for each standard. The concentration of metabolite, in pg, of each sample is then obtained from the standard curve. A typical standard curve is shown in Fig. 2.

2.3 Assay Validation

The NSB of a properly set RIA procedure should be between 1 and 5%. Also it is important to validate an RIA measurement by determining its linearity, reproducibility and specificity.

Linearity This test determines whether the trend in detection of metabolites by the assay is linear. The linearity of an assay must be determined in same media where sample is to be tested. To test the linearity of an assay, different amounts of the PG are added to the sample medium, plasma, urine or buffer. The amount added must be within the range of assay procedure, i.e. 0–200 pg.

An assay is done as described above. Then the calculated amount of metabolite in media is plotted against the amount (in pg) added to the media. The plot should yield a straight line with a slope of 1. If the slope varies from 1, then the media contains some interfering compounds and an extraction should be done to remove them.

Assay duplication This test is meant to ensure that the results obtained from diluted and undiluted samples are equivalent. The test is done as in the linearity test except that the standard is diluted, as an example, 1:2, 1:4, 1:10 etc. Calculated results are plotted against the standard curve. The resulting plot should parallel the standard curve. The variability of diluted and undiluted standards should be between 5 and 10%.

This is a measure of the precision of an assay and also a test of the consistency and reproducibility of a technique. It is measured by the inter-assay (results between one assay and another) and intra-assay (results between one sample, for instance, several times within an assay) variability. The inter-assay and intra-assay variabilities should be less than 10%.

Assay variability

2.4   Example: Measurement of PGs and TxA$_2$ Production by Foetal Rabbit Lungs

We measured prostacyclin, PGI$_2$, PGE$_2$ and TxA$_2$ as an index of arachidonic acid turnover in isolated lungs of foetal New Zealand White (NZW) rabbits (gestational age 28 days).

Materials

- Krebs bicarbonate buffer (119 mM NaCl, 4.7 mM KCl, 1.2 mM MgSO$_4$, 22.6 mM NaHCO$_3$, 1.2 mM KH$_2$PO$_4$, 1.6 mM CaCl$_2$, and 5.5 mM glucose; pH 7.4)
- Antisera for 6-keto-PGF$_{1\alpha}$, PGE$_2$ and TxB$_2$ (PerSeptive Diagnostics, Cambridge MA)
- Eicosanoid standards (Biomol Research Laboratories, Plymouth Meeting, PA)
- [^{3}H]-PG tracers (DuPont NEN, Boston MA)
- Liquid scintillation cocktail (ICN Biochemicals, Irvine, CA)

Procedure

1. Isolate lungs of foetal NZW rabbits. Rinse gently with 5 ml Krebs' buffer.

2. Transfer each washed lung into a 16 × 100-mm glass culture tube containing 2 ml of buffer equilibrated at 37 °C in the bath shaker and aerate with the gas mixture 21% O$_2$, 5% CO$_2$, rest N$_2$.

3. Challenge each lung with 1 µM of calcium ionophore. A23187 and allow to incubate for 10 min.

4. After 10 min incubation, remove the lung tissue and extract the media for PGs as described.

5. Assay PGI_2 as its stable metabolite 6-keto-prostaglandin $F_{1\alpha}$ and TxA_2 as its stable metabolite TxB_2.

The cross-reactivity of each antiserum with other eicosanoids is <0.01%. RIA measurements are based on standard curves using an authentic standard of each metabolite in a concentration range of 0–200 pg/0.3 ml for 6-keto-$PGF_{1\alpha}$ TxB_2 and PGE_2. The 6-Keto-$PGF_{1\alpha}$ and PGE_2 antisera produced 50% displacement of bound standard at 25 pg. Curve fitting of standard and calculation of amount of metabolite is done with a weighted non-linear RIA programme. Final results are then corrected for recovery and expressed as ng/g dry lung weight. The results are shown in Fig. 3. Lungs of foetal NZW rabbit of 28 days gestation metabolised AA into PGI_2, TxA_2, and PGE_2.

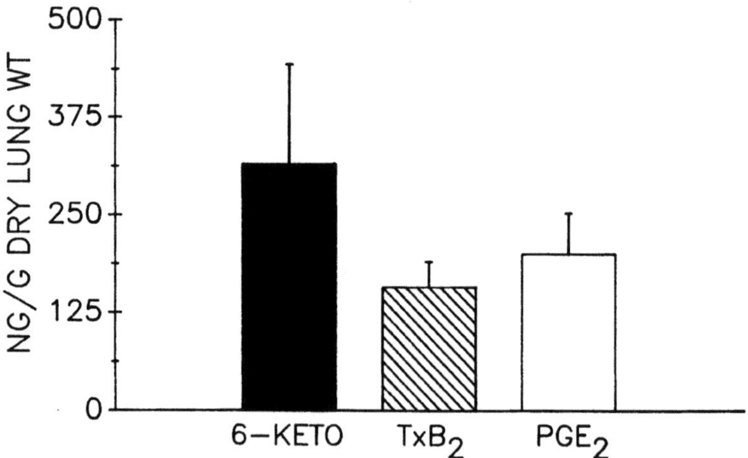

Fig. 3. Prostacylin, TxA_2 and PGE_2 measured from incubated lungs of 28 days gestation foetal NZW rabbits. Prostacylin was measured as 6-keto-$PGF_{1\alpha}$ while TxA_2 was measured as TxB_2. Data are means ± SEM of seven experiments

Radioimmunoassay buffers for PGs and TxA_2 are different. Use **Modifications** of potassium phosphate/EDTA (PET/BSA) buffer results in a better binding of PGE_2 and 6-keto $PGF_{1\alpha}'$ antisera. The PET/BSA buffer contains 3.8 mM KH_2PO_4, 1.0 mM EDTA, 0.1% BSA, and 0.01% thimerosal. TxB_2 assay is done with a phosphate buffered saline (PBS) containing 100 mM NaH_2PO_4, 151 mM Nacl, 0.1% polyvinyl- pyrrolidone (Sigma Chemicals, #PVP-40), and 0.1% sodium azide.

The dextran-coated charcoal for the PBS and the PET/BSA buffers contains (per 100 ml of buffer) 125 mg Dextran T-70 (Sigma Chemicals, #D-1390) and 1.25 gm RIA grade charcoal (Sigma Chemicals #C-5385). The charcoal suspension is prepared by first dissolving the dextran in the buffer before the charcoal is added. Manual agitation of the charcoal suspension is preferred.

2.5 Tips, Tricks and Troubleshooting

• The Standard Curve

The RIA standard curve presented in Fig. 2 may be viewed as having three distinct regions. The first region is an area of steep decline in percentage bound from 0–50 pg. This is the most sensitive portion of the curve. It is recommended that sample dilutions should be adjusted to fall within this sensitive area of the curve since very small change in sample concentration can easily be detected.

The second region, from 50–100 pg, is less sensitive because of the gradual drop in percentage bound with an increase in concentration of the standard used. However, the region is still sensitive enough to detect small changes in sample concentration.

The third region, from 100 pg tends to be asymptotic with the concentration axis. Small changes in sample concentration may not be detected with great confidence in this region.

• General problems in RIA

Some general problems may arise in the process of RIA measurements. Below are given some recurrent problems of RIA and their possible solutions.

• Slope of Curve is Flat, but 0 pg Binding is Normal

This is caused by use of degraded or wrong eicosanoid standards. If this occurs, a new standard should be made or the correct standard should be used. It is important not to assume that the correct standard has been used. A physical examination of the vial containing the standard should be made to make sure of its condition.

• Slope of Curve is Flat, but 0 pg Binding is High

This means that too much antibody is used in the assay. The problem is solved by increasing the dilution of the antibody.

• None or Low pg Binding

This suggests the following:

– Wrong or contaminated assay buffer; then use fresh buffer.
– No antibody was added; redo the assay with correct antibody dilution.
– Antibody or [^{3}H]-tracer is degraded, then make fresh antibody or tracer dilution.

• A High or Low NSB

If NSB is >5% the charcoal suspension is old or incorrectly made, make fresh charcoal suspension correctly. A high NSB can also mean that the [^{3}H]-tracer is degraded. If the NSB is <1%, too much charcoal is used, therefore reduce the amount of charcoal.

3 Measurement of Eicosanoids by High Performance Liquid Chromatography (HPLC)

High performance liquid chromatography utilising either UV or fluorescence detector is another method of measuring lipid turnover in cells and tissues (Bakhle et al. 1985; Ibe et al. 1991). Prostaglandins and the other cyclooxygenase metabolites of AA can also be measured by HPLC. However, use of HPLC to measure PG and its metabolites is limited because these eicosanoids do not absorb in the UV range and must be chemically modified before UV or fluorescence detection can be achieved. If radiolabelled AA is used in the study, HPLC assay offers an excellent method to measure AA turnover (Capdevila et al. 1981).

Leukotrienes and HETES and the other group of eicosanoids are very amenable to HPLC quantitation. Leukotrienes are conjugated trienes and are measured by HPLC at 280 nm while HETES are conjugated dienes and are measured at 235 nm. In this section, measurement of LTs by HPLC will be described. Samples for HPLC analysis must be extracted and concentrated in order to be accommodated by the extremely small volume of samples required in HPLC analysis.

3.1 Extraction of Samples for HPLC Analysis

The preferred method used to extract LTs for HPLC assay utilises a solid support as described below (Ibe and Campbell 1988).

Materials

- Sample media for LTs measurement
- Disposable C_{18} column cartridges 6 ml volume (J.T. Baker, #7020–6), Low pressures extraction manifold (Sep-Pak cartridge rack, Waters Associates)
- PGB_2 for use as internal standard

- Methanol
- 0.1% EDTA
- Deionised distilled water
- 13 × 100-mm glass vials pre-rinsed with methanol

Procedure

1. Add to each sample 250 ng internal standard PGB_2 and known amount of a [³H]-leukotriene standard, usually 3000 cpm of [³H]-LTC_4, to each unknown to determine recovery of sample from the extraction process.

2. Place the 6 ml C_{18} column, one for each sample, in the Sep-Pak cartridge rack and wash column with 5 ml of methanol, followed by 5 ml of water and 5 ml of 0.1% EDTA.

3. Place sample on extraction column and load in column by pulling with low vacuum. Wash column with 5 ml of water. The column should not run dry. Eluates of steps 2 and 3 should be discarded.

4. Elute LTs with 5 ml of methanol. Collect sample eluates in the methanol rinsed 13 × 100 mm glass tubes.

5. Evaporate methanol under nitrogen at 35–40 °C. Reconstitute sample residue in 0.3 ml of 70% methanol: water. Determine sample recovery from extraction by counting 0.01 ml aliquot of the resuspension. Recovery of LTs is usually 80–86%. Centrifuge the reconstituted suspension at 3000 rpm for 10 min on regular IEC bench centrifuge to remove solid particles that may clog HPLC column. The supernatant is loaded on to HPLC according to the capacity of the sample loop.

! **Note.** The column can be reused by washing with 10 ml of methanol. Each column can be reused five to ten times. However, the extraction efficiency decreases with each reuse. With plasma, microsomes, cytosol or tissue homogenate samples, columns can only be reused two to three times before they become heavily clogged up. Therefore, for samples in these types of media, one column per sample is recommended.

3.2   HPLC of Leukotrienes

To measure LTs (LTB_4, LTC_4, LTD_4 and LTE_4) by HPLC, a standard calibration curve is set up from which the amount of LTs in the sample is determined. One important point to remember is to be sure that the standards to be used in the calibration curve are treated in a similar manner as the samples. Also the standards used must be those of the LT or LTs to be measured. HPLC analysis of LTs employs either a normal phase (NP) or a reversed phase (RP) HPLC column. NP-HPLC uses lipophilic solvents such as hexane or petroleum ether to elute samples from the HPLC column. RP-HPLC uses hydrophilic solvents such as methanol and acetonitrile to elute the samples.

- HPLC equipment with UV detector and integrator
- NP- or RP-HPLC column with a matching guard column
- Sample in 70% methanol
- LTs standards (commercially available)
- Solvent (mobile phase) for elution of appropriate composition to afford a timely elution of metabolites from column

Materials

1. Choose the appropriate composition of mobile phase to elute metabolites from column. The choice of mobile phase composition is easily obtained from the literature on LTs assay.

2. Equilibrate HPLC column with mobile phase for 15 min until a stable baseline is obtained.

3. If using column and/or mobile phase for the first time, run standards individually to authenticate the retention times (T_r).

4. Load and inject three to four different concentrations of mixture of LTs standards for use in generating the standard calibration curve. The samples are run after the standards.

HPLC procedure

5. Elute metabolites with the mobile phase.

! **Note.** Do not forget to turn on the integrator.

6. After last run of the day, generate standard curve by plotting peak area ratio (PAR) of each standard to the internal standard.

7. Compute amount of LTs in each sample from the standard curve by interpolating the PAR of metabolite present in sample.

8. After each day's run, column should be washed with 10x its volume and stored in the organic solvent.

9. Steps 2–8 should be repeated each day when an assay is to be done.

3.3 Assay Validation

Just as with RIA, it is important to validate an HPLC measurement by determining the linearity, reproducibility and specificity of the assay method.

Linearity This test determines whether the metabolites detected by the calibration curve falls within the linear range of the curve. The linearity of an assay must be determined in the same media in which sample is to be tested. To test the linearity of an assay, different amounts of LTs are added to the sample medium (plasma, urine or buffer), and extracted. The amount added must be within the range of standard curve. An assay is then done as described above. Then the calculated amount of metabolite in media is plotted against the amount (in pg) added to the media. The plot should yield a straight line with a slope of 1. If the slope varies significantly then the precision of extraction and pipetting should be improved.

This is done as described for RIA (see Sect. 2.3). The duplicate **Assay**
values should not vary by more than 10%.   **duplication**

This is a measure of the precision of an assay and also tests the **Assay**
consistency and reproducibility of a technique. It is measured **variability**
by the inter-assay (results between one assay and another) and
intra-assay (results of one sample, for instance, several times
within an assay) variability. The inter-assay and intra-assay
variabilities should be less than 10%.

3.4 Example: HPLC Measurement of LTs Produced by Foetal Rabbit Lungs

We measured LTs (LTC_4, LTD_4, LTE_4 and LTB_4) production as
an indicative of arachidonic acid turnover in isolated lungs of
foetal NZW rabbit (gestational age 28 days).

- Krebs' bicarbonate buffer: 119 mM Nacl   **Materials**
  4.7 mM KCl
  1.2 mM $MgSO_4$
  22.6 mM $NaHCO_3$
  1.2 mM KH_2PO_4
  1.6 mM $CaCl_2$
  5.5 mM glucose
  pH 7.4
- Eicosanoid standards (Biomol Research Laboratories, Plymouth Meeting, PA)

1. Isolate lungs of foetal NZW rabbits and rinse them gently **Procedure**
   with 5 ml of the Krebs' buffer.

2. Transfer each washed lung into a 16×100-mm glass culture
   tube containing 2 ml of buffer, equilibrated at 37 °C in the
   bath shaker and aerated with the gas mixture 21% O_2, 5%
   CO_2 and rest N_2.

3. Challenge each lung with $1\,\mu M$ of calcium ionophore A23187 and incubate for 10 min.

4. After 10 min of incubation, remove the lung tissue and extract the media for LTs as described and quantitate on a RP-HPLC (Ibe et al. 1991).

! **Note.** The HPLC equipment was an ISCO model 2360 pump with a valco injector, model 2350 ternary gradient programmer and a V^4 variable wavelength detector (Instrument Specialists Co., Lincoln, Neb). The detector was connected to an HP3394 integrator (Hewlett-Packard, Palo Alto, Ca). The RP-HPLC column was a Spherisorb C_{18}, $5\,\mu m$, $4.6 \times 250\,mm$ (Instrument Specialists Co., Lincoln, Neb) with a $10\,\mu m$ C_{18} guard column (Phenomenex, Rancho Palos Verdes, Ca). Mobile phase was a mixture of methanol: water: glacial acetic acid (50:50:0.1) buffered at pH 5.6 with ammonium hydroxide.

Elution of metabolites was done at a flow rate of 1 ml/min beginning with 50% methanol for 5 min, then changed to 58% methanol isocratically for 45 min. Retention times of the leukotrienes were: LTC_4, 17 min; PGB_2, 21 min; LTD_4, 32 min; LTB_4, 39 min; LTE_4, 41 min.

The calibration curve should be linear at A_{280} for each metabolite tested. The linearity is tested by comparing the slope and correlation coefficient of calibration curves of added standards to the amount calculated from area under the chromatographic peaks.

The slope of each metabolite was as follows (slope, correlation coefficient): LTC_4, 1.08 ± 0.04, 0.98; LTD_4, 0.84 ± 0.03, 0.85; LTE_4, 0.94 ± 0.01, 0.88; LTB_4, 1.01 ± 0.02, 0.95. The concentration of leukotriene metabolites in the sample is obtained from a calibration curve using authentic leukotrienes that is run on each day of analysis. The results of a typical assay are shown in Fig. 4. Foetal NZW rabbit of 28 days gestation metabolised AA acid into LTC_4, LTD_4, LTE_4, and LTB_4, which were measurable by HPLC method.

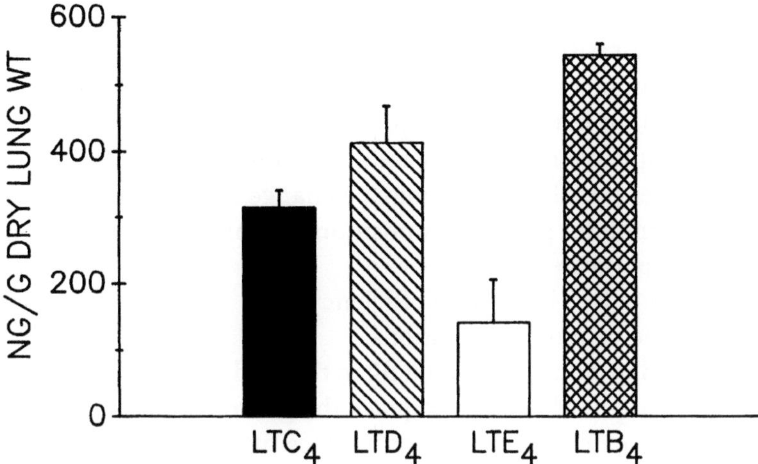

Fig. 4. Profile of leukotrienes synthesised by lungs of 28 days gestation foetal NZW rabbits. Data are mean ± SEM of seven experiments

3.5   Tips, Tricks and Troubleshooting

- It is important to run a calibration curve on each day of an analysis. HPLC buffer should be made fresh daily. The organic solvent, methanol or acetonitrile should be degassed before use. Composition of mobile phase should be chosen such that an assay is completed within a minimum length of time and good resolution of peaks is obtained.

  HPLC troubleshooting. Some problems may arise in the process of HPLC analysis. The following are some problems that often arise during an assay.
- Problem: unrealistic high system back pressure
  Possible causes and solutions:
  - Clogged inlet frit; back-flush column to dislodge clog or replace frit.
  - Clogged in-line filter; replace filter.
  - Clogged guard column; replace guard column.
  - Clogged tubing, injector, or flow cell; replace tubing, flush/replace injector and/or flow cell.

- Problem: unexpected change in retention times
  Possible causes and solutions:
  - Column contamination; clean or replace column.
  - Column inadequately equilibrated; equilibrate column.
  - Sample overload; decrease amount of sample injected.
  - Other possible causes may be change in flow rate, change in mobile phase composition or pH and change in temperature.
- Problem: loss in column efficiency
  Possible causes and solutions:
  - Void in column packing; fill in the void or replace column.
  - Spent guard column; replace the guard column.
  - Improper column fittings; replace fittings.
- Problem: peak tailing or splitting
  Possible causes and solutions:
  - Dirty column; clean or replace column.
  - Particulates on inlet frit; clean or replace frit.
  - Loss of bonded phase or spent guard column; replace column.
  - Change in pH; make fresh buffer.
- Problem: mystery peaks
  Possible causes and solutions:
  - Contaminated column; clean or replace column.
  - Air bubbles in assay; degas solvent properly.
  - Electronic problems; replace the problematic parts.

4 Summary

Although an HPLC method for quantifying leukotrienes has been described, it is necessary to state that RIA methods are also available to measure individual LTs. An advantage for the HPLC method for measuring LTs is that the four leukotrienes of interest, viz. LTB_4, LTC_4, LTD_4 and LTE_4 can be measured in one HPLC run. If an RIA of LTs is preferred, the antibody can be purchased from commercial sources such as Caymen Chemicals, PerSeptive Diagnostics or Biomol Research Laboratories.

References

Bakhle YS, Moncada S, deNucci G, Salmon JA (1985) Differential release of eicosanoids by bradykinin, arachidonic acid and calcium ionophore A23187 in guinea-pig isolated perfused lung. Br J Pharmacol 86:55–62

Campbell WB, Ojeda SR (1987) Measurement of prostaglandins by radioimmunoassay. Methods Enzymol 141:324–341

Capdevila J, Chacos N, Werringloer J, Prough RA, Estabrook RW (1981) Liver microsomal cytochrome P-450 and oxidative metabolism of arachidonic acid. Proc Natl Acad Sci USA 78:5362–5366

Fitzpatrick FA, Gorman RR, Wynalda MA (1977a) Electron capture gas chromatographic detection of thromboxane B_2. Prostaglandins 13:201–208

Fitzpatrick FA, Wynalda MA, Kaiser DG (1977b) Oximes for high performance liquid and electron capture gas chromatography of prostaglandins and thromboxane. Anal Chem 49:1032–1035

Hassid A, Levine L (1977) Stimulation of Phospholipase activity and prostaglandin biosynthesis by melittin in cell culture and in vivo. Res Commun Chem Pathol Pharmacol 18:3400–3404

Henderson WR Jr (1987) Eicosanoids and lung inflammation. Am Rev Respir Dis 135:1176–1185

Holtzman MJ (1991) Arachidonic acid metabolism. Implications of biological chemistry for lung function and disease. Am Rev Respir Dis 143:188–203

Huber M, Kastner S, Scholmerich J, Gerok W, Keppler D (1989) Analysis of cysteinyl leukotrienes in human urine: enhanced excretion in patients with liver cirrhosis and hepatorenal syndrome. Eur J Clin Invest 19:53–60

Ibe BO, Campbell WB (1988) Synthesis and metabolism of leukotrienes by human endothelial cells: Influence on prostacyclin release. Biochim Biophys Acta 960:309–321

Ibe BO, Isenberg WB, Raj JU (1991) Endogenous arachidonic acid metabolism by calcium ionophore A23187-stimulated lamb lungs: effect of hypoxia. Am J Respir Cell Mol Biol 4:379–385

Powell WS (1980) Rapid extraction of oxygenated metabolites of arachidonic acid from biological samples using octadecylsilyl silica. Prostaglandins 20:947–957

Strife RJ, Murphy RC (1984) Stable isotope labelled 5-lipoxygenase metabolite of arachidonic acid: analysis by negative ion chemical ionisation mass spectrometry. Prostaglandins Leukotrienes Med 13:1–8

Turk J, Weis SJ, Davis JE, Needleman P (1978) Fluorescent derivative of prostaglandins and thromboxane for liquid chromatography. Prostaglandins 16:291–309

Willerson JT, Golino P, Edit J, Campbell WB, Buja LM (1989) Specific platelet mediators and unstable coronary artery lesions: experimental evidence and potential clinical implications. Circulation 80:198–205

Chapter VIII Lipids and Signal Transduction

P.S. Sastry

1 Background

Cells of multicellular organisms constantly receive signals from other cells and from the environment which they perceive, interpret and to which they respond with an appropriate metabolic or physiological change. These signals may be either physical, such as light and sound, or chemical, such as growth factors, hormones and neurotransmitters. They interact with specific molecules in the cell's plasma membrane called receptors, and through a cascade of events ultimately produce a physiological change. This cascade, known as signal transduction, involves, besides the receptors, a group of GTP-binding proteins – the G proteins, some membrane-bound enzymes and their substrates which produce a number of small molecules within the cell, called the second messengers. These, in turn, through their effects on protein kinases or phosphatases, modulate the phosphorylation status of a variety of intracellular proteins and eventually result in cellular processes such as metabolism, excitation, secretion, contraction, sensory perception, gene expression and cell growth.

The complexity of the signal transduction mechanism achieves the twin objectives of severalfold signal amplification and a high degree of specificity in cellular response. It has been known for some time that many receptors utilise a signal path-

Department of Biochemistry, Indian Institute of Science, Bangalore-560012, India

way in which the enzyme adenylate cyclase present on the cytoplasmic side of the membrane is activated through G proteins to form the second messenger – cyclic AMP, intracellularly. In the recent past, however, it has been discovered that some membrane-bound phospholipids play a vital role in signal transduction mechanisms of a large number of receptors. In fact, the lipid-mediated signal pathways now appear to be more ubiquitous and versatile.

At present, at least four phospholipid classes have been shown to participate in signal transduction. These are: phosphatidylinositols, phosphatidylcholines, sphingolipids, and glycosylphosphatidyl-inositols. These mechanisms are briefly reviewed and the relevant methodology is described here.

2 Phosphatidylinositol Signal Pathway

Phosphatidylinositols account for about 6–8% of the total phospholipids in animal cells and comprise: phosphatidylinositol [1-(3-sn-phosphatidyl)-D-myoinositol] (PI), phosphatidylinositol-4-phosphate (PIP) and phosphatidylinositol 4,5-bisphosphate (PIP$_2$); (Fig. 1). Of these, PI accounts for over 90%, and most of it is in the endoplasmic reticulum (ER). In contrast, most of PIP, PIP$_2$ and only a small amount of PI are in the plasma membrane. Their diglyceride moiety is mainly 1-stearoyl-2-arachidonoyl-sn-glycerol and these phospholipids are metabolically very active.

More than three decades ago, it was demonstrated that agonist activation of membrane receptors stimulates PI turnover, and since then this phenomenon was observed with several agonists (For a list of agonists, see Chandrasekhar and Hokin 1986). However, only recently the mechanistic details of this signal pathway were established.

This pathway involves agonist-stimulated phosphodiesteratic (phospholipase C) cleavage of PIP$_2$ to generate the

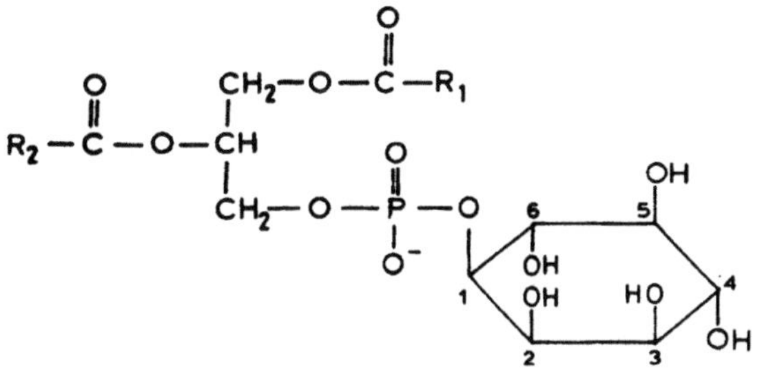

1 – (3 – sn – Phosphatidyl) – D – myoinositol (PI)

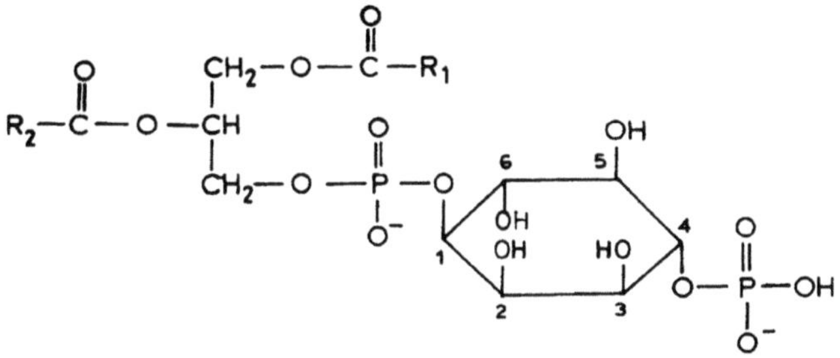

1 – (3 – sn – Phosphatidyl) – D – myoinositol – 4 – Phosphate (PIP)

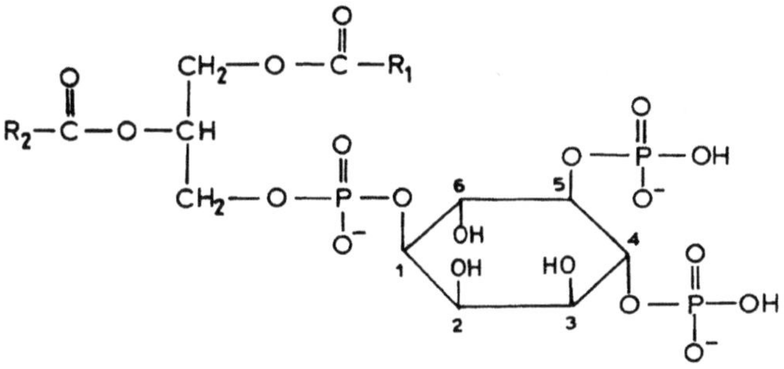

1 – (3 – sn – Phosphatidyl) – D – myoinositol 4,5 – bisphosphate (PIP_2)

Fig. 1. Structures of phosphatidylinositols

second messengers: inositol-1,4,5-trisphosphate [Ins-(1,4,5)-P_3] and diacylglycerol (DG). Ins-(1,4,5)-P_3, through its specific receptor present on ER, releases Ca^{2+} from ER, elevates intracellular Ca^{2+} which activates calmodulin-dependent protein phosphorylation. The DG activates protein kinase C and phosphorylates a different set of proteins. This change in the phosphorylation status of intracellular proteins results in physiological changes. Increased levels of both Ins-(1,4,5)-P$_3$ and DG appear to be necessary for maximal responses.

The elevated level of intracellular Ca^{2+} may also activate some calmodulin-independent kinases. In yet another reaction, arachidonate is formed by either phospholipase A$_2$ action on phosphoinositides or by lipase action on DG. Arachidonate is the obligatory substrate for the synthesis of eicosanoids, which mediate a variety of autocrine functions. Thus, the phosphoinositide signal pathway has the potential to generate a wide variety of second messengers depending on the agonist, receptor and cell type, and to bring about specific physiological changes (Fig. 2).

The levels of PI, PIP and PIP$_2$ in the membranes are in constant flux through the activity of PI synthase, PI kinase and PIP kinase. Though Ins-(1,4,5)-P$_3$ appears to be the first product in this signal pathway, it is immediately metabolised to a complex set of inositol bisphosphates, monophosphates and also tetrakis-, pentakis- and hexakis-phosphates, the physiological significance of which is not yet understood. Lithium amplifies the accumulation of inositol phosphates through its inhibitory action on inositol-1-phosphate and inositol polyphosphate phosphatases.

2.1   Isolation and Separation of Phosphoinositides

Among the phosphoinositides, PI along with other phospholipids can be extracted from animal tissues with neutral organic solvents. However, the polyphosphoinositides, PIP and PIP$_2$, can be extracted only with acidic solvents. Moreover, the polyphosphoinositides are degraded very rapidly post mortem, necessitating immediate cooling of the tissue. Keeping these requirements in view, an extraction procedure origi-

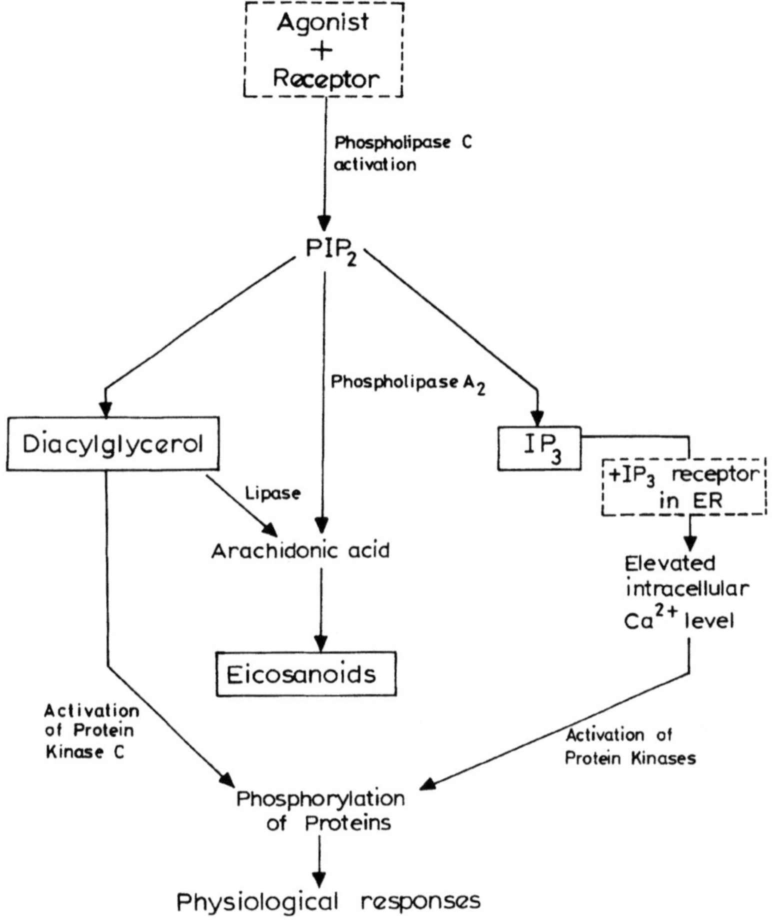

Fig. 2. Phosphatidylinositol signal pathway and generation of multiple second messengers

nally reported by Dawson and Eichberg (1965) for brain tissue, which is richest in these lipids, is described below:

2.1.1 Extraction of Phosphoinositides

- Rat brain tissue   **Materials**

- Reagents
  - Hydrochloric acid
  - Chloroform (AR)
  - Methanol (AR)
  - Liquid nitrogen
  - 0.9% Sodium chloride solution

Equipment
- Waring blender
- Mortar and pestle (medium sized)
- Ice bath (to cool solvents)
- Corex tubes (Sorvall, 30 ml)
- Rotary evaporator
- Vortex mixer
- Sorvall-5B Centrifuge with SS-34 Rotor

Extraction of phospho-inositides

1. Sacrifice 8–10 young albino rats (weighing about 120 g) by cervical dislocation, decapitate and quickly dissect out the whole brain (cerebral hemispheres and cerebellum). Collect approximately 10 g of tissue and put immediately in liquid nitrogen (about 50 ml) in a medium sized mortar. Grind the tissue into a powder with a pestle by adding small amounts of liquid nitrogen, if necessary.

2. Transfer the ground tissue into a high-speed blender with ice-cold chloroform:methanol (1:1, v/v) using a total of 100 ml of solvent (10 times of the brain weight). Homogenise the tissue for 1 min.

3. Transfer equal amounts of the homogenate into 4×30-ml Corex tubes and centrifuge at 9000 rpm for 5 min in a Sorvall RC-5B at 0–5 °C. Decant the supernatant and save.

4. Resuspend the pellets in 30 ml of chloroform:methanol (2:1, v/v) per tube. Mix thoroughly with a vortex mixer and centrifuge at 9000 rpm for 5 min. Decant the supernatant and mix with the supernatant obtained at step 3. Set aside the pellets on ice for extraction of polyphosphoinositides.

5. To the combined supernatants (about 220 ml) add 50 ml of chloroform to make the final chloroform:methanol ratio (2:1, v/v). Transfer it to a 500-ml separating funnel, add 54 ml (0.2 volumes) of 0.9% sodium chloride solution, mix thoroughly and allow the phases to separate for 3 to 4h. After the phases separate completely, leaving only a thin protein-aqueous interface, collect the lower chloroform-rich phase. Evaporate the solvent to near dryness in a rotary evaporator (see Chap. III, this Vol.), dissolve in chloroform (use a few drops of methanol if necessary to solubilise) and transfer to a stoppered 50-ml volumetric flask and store at −20 °C. This extract will contain most of the lipids including PI. Most animal brain tissues contain about 5% of phospholipids. Therefore, from 10 g, about 500 mg of phospholipids should be expected. Of this about 3% (i.e. 15 mg) would be the PI content. PI can be separated from other lipids by thin layer chromatography (as described in Sects. 2.1.2 and 2.1.3) and estimated.

6. To the pellets obtained in step 4, add 10 ml of chloroform:methanol (2:1, v/v) containing 0.25% of concentrated HCl and incubate for 20 min at 37 °C. Centrifuge the tubes at 9000 rpm for 5 min in Sorvall RC-5B. Carefully decant the supernatant into a 250-ml conical flask through a funnel plugged with glass wool to prevent transfer of tissue particles.

7. To each pellet add 10 ml of chloroform:methanol (2:1, v/v) containing 0.25% concentrated HCl, mix thoroughly, incubate, centrifuge and collect supernatant as in step 6.

8. Repeat step 7.

9. Take the combined supernatants (about 120 ml) in a 250-ml separating funnel, add 24 ml (0.2 Vol.) of 1N HCL, shake well and allow the phases to separate in about 2 to 3h. Collect the lower chloroform phase in a conical flask and discard the upper phase.

10. Transfer the chloroform phase to a 250-ml separating funnel, add 0.5 Vol. (about 40 ml) of chloroform:methanol:1N HCl (3:48:47, v/v). shake gently and allow the phases to separate.

11. Collect the lower chloroform phase in a 500-ml round bottomed flask, take to dryness in a rotary evaporator and dissolve the solids in chloroform and store in 25-ml volumetric flask at −20 °C. This extract will contain mainly PIP and PIP$_2$ with traces of PI. From 10 g of rat brain, about 2–3 mg of PIP and 10 mg of PIP$_2$ should be expected.

2.1.2 Separation of PI by Two-Dimensional TLC

The extract obtained at step 5 in the procedure for extraction of phosphoinositides (Sect. 2.1.1) contains PI along with other lipids. PI can be separated from other lipids by TLC (Lagos and Vergara 1990).

Materials
- 10 × 10-cm high efficiency TLC plates (Analtech HETLC-HL, Newark, DE, USA).
- Chromatography jars
- Solvent mixtures:
  a) Chloroform:methanol:28% aqueous ammonia (65:35:5, v/v)
  b) Chloroform:methanol:acetone:glacial acetic acid:water (60:12:24:12:6, v/v)

Separation by two dimensional TLC

1. Activate the TLC plates at 140 °C for 1 h and cool in a desiccator.

2. Line two chromatography jars with Whatman #3 filter paper. Add solvent (a) in one jar and solvent (b) in the other. Close the jars and allow them to saturate with the solvent for 1 h.

3. Apply 50 to 100 µl of the extract as a small spot at the origin.

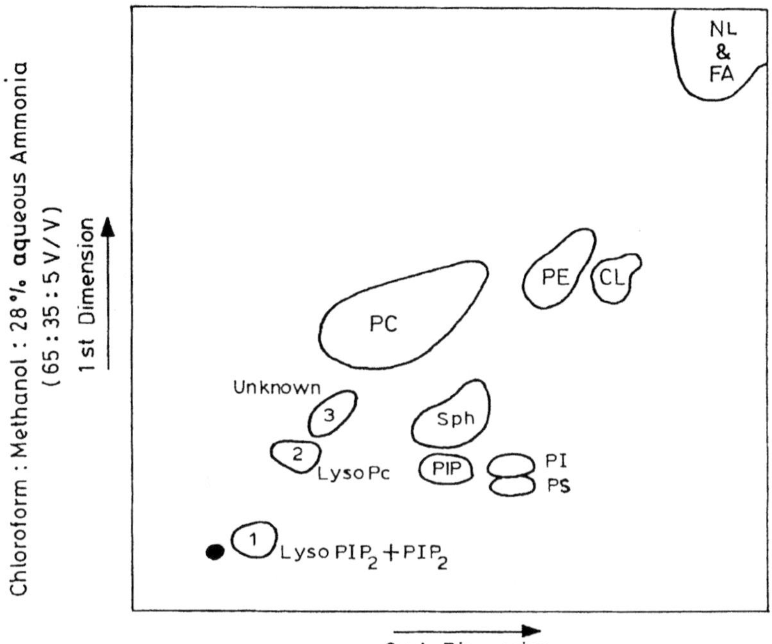

Fig. 3. Separation of lipids by two-dimensional TLC; *PI* phosphatidylinositol; *PIP* phosphatidylinositol-4-phosphate; *PIP₂* phosphatidylinositol-4,5-bisphosphate; *PC* phosphatidylcholine; *PE* phosphatidylethanolamine; *PS* phosphatidylserine; *Sph* sphingomyelin; *CL* cardiolipin; *NL* neutral lipids; *FA* fatty acids

4. Develop the TLC in solvent (a) until the solvent reaches the top (1st direction). Remove the plate and let it dry for about 10 min.

5. Turn the plate by 90° and develop the TLC with the solvent (b) in the other direction until the solvent reaches the top. Remove the plate and let it dry for 10 min.

6. Expose the plate to iodine vapour and mark the areas. A typical separation of the lipids on TLC is shown in Fig. 3.

! **Note.** If required, the quantity of PI can be determined by estimating the amount of phosphorus in the PI spot with appropriate controls as described in Chapter IV, this Volume.

2.1.3 Separation of PI by One-Dimensional TLC

The extract obtained at step 11 (Sect. 2.1.1) contains essentially PIP, PIP_2 and a small amount of PI. These can be separated by TLC on silica gel H plates impregnated with oxalate.

Materials
- Chromatography jars
- Desaga applicator
- 20 × 20-cm glass plates
- Silica gel H plates impregnated with 1% oxalate.

! **Note.** Suspend 40 g of silica gel H (Acme Synthetic India) or from any other specific source in 100 ml of 1% potassium oxalate by thorough mixing and spread it on four 20 × 20-cm glass plates with the help of a Desaga applicator to give a uniform thickness of 0.5 mm and allow to dry.

- PIP and PIP_2 as standards (Sigma Chemicals, USA)
- Solvent mixture: chloroform:methanol:conc. ammonium hydroxide:water (40:48:5:10, v/v).

Separation by one-dimensional TLC

1. Activate the TLC plates at 120°C for 3 h and cool in a desiccator.

2. Line the chromatography jar with Whatman #3 filter paper. Add the solvent to the jar and close. Allow the jar to saturate with the solvent for 1 h.

3. Apply 50–100 µl of the extract as a small spot at the origin. Simultaneously, apply commercially obtained PIP and PIP_2 as separate spots for comparison.

4. Develop the plate with the solvent until it reaches the top. Remove the plate and let it dry for 10 min.

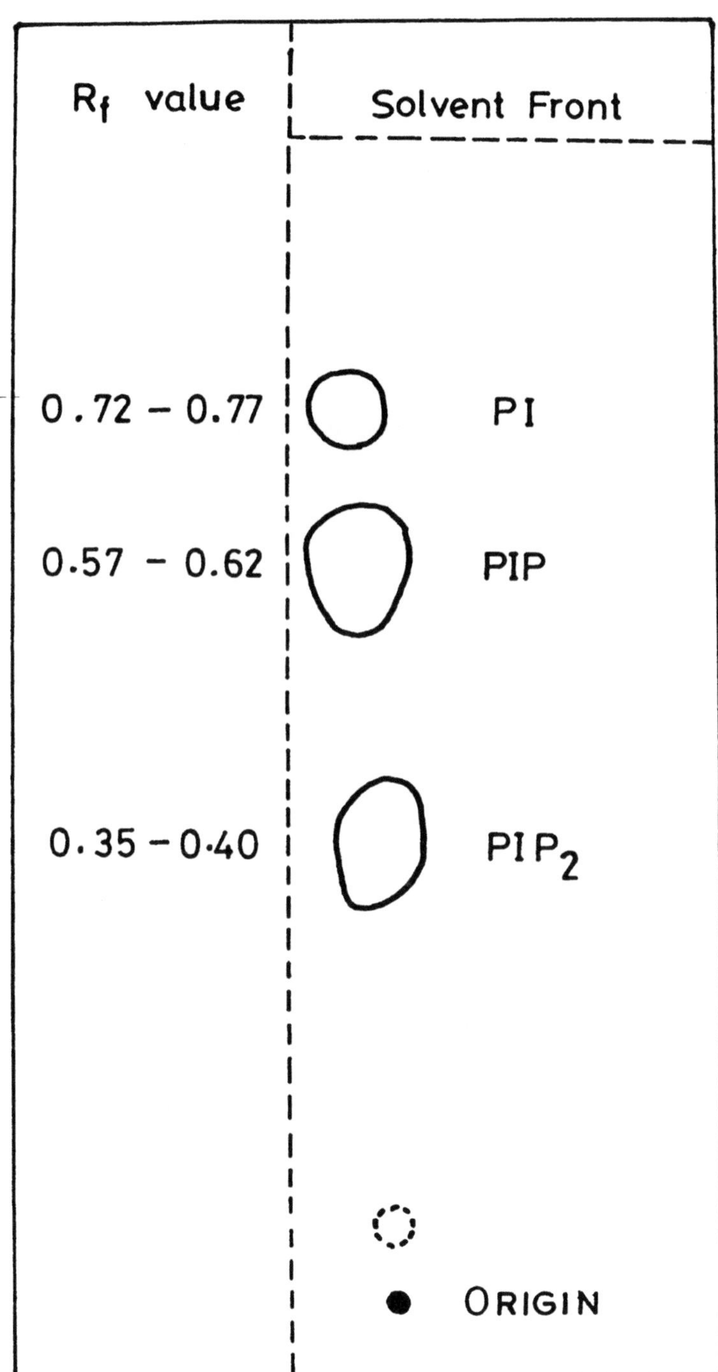

Fig. 4. Separation of polyphosphoinositides by one-dimensional TLC. ● origin; *PI* phosphatidylinositol; *PIP* phosphatidylinositol 4-phosphate; *PIP$_2$* phosphatidylinositol-4,5-bisphosphate. Solvent: chlorofom:methanol:conc. ammonium hydroxide:water (40:48:5:10, v/v)

5. Expose the plate to iodine vapour. A typical separation of PI, PIP and PIP_2 is shown in Fig. 4.

! **Note.** The contents of PI, PIP and PIP_2 can be quantified by the determination of their phosphorous content with appropriate controls as detailed in Chapter IV, this Volume.

2.2 Assay for Agonist-Stimulated Formation of Inositolphosphates (IPs)

The signal transduction mechanism for several receptors involves the hydrolysis of phosphoinositides via the activation of phospholipase C and production of IPs. This phenomenon can be studied by measuring the formation of IPs on stimulation of [^{3}H]-inositol prelabelled tissue with appropriate agonists in presence of Li^+. In most studies, the magnitude of stimulation is measured by assaying the total amount of all IPs formed (Blackstone et al. 1989).

However, since several isomers of inositol mono- and polyphosphates are formed when tissues are stimulated with agonists, it is of great interest to separate and measure their individual quantities. This can be achieved by separating the various inositol phosphates by high pressure liquid chromatography (HPLC) and determining the radioactivity associated with each of them (Sastry et al. 1992).

A method for the specific determination of Ins-(1,4,5)-P_3, the first product of phospholipase C action on PIP_2, is also available. This is a receptor binding assay which determines the mass of Ins-(1,4,5)-P_3 (Challis et al. 1990). These methods for experiments with brain tissue are described in Section 2.2.1, but can be adapted for other tissues with appropriate agonists.

2.2.1 Assay for Total IPs Formed on Agonist Stimulation

- Male rats (weighing 150–200 g)   **Materials**
- *myo*-[2-³H]-Inositol (specific activity 15–17 Ci/mmol) from New England Nuclear Research Products (Boston, MA, USA)
- Carbachol, Dowex AG 1 × 8 resin (200–400 mesh) formate form (Sigma Chemicals, USA)
- 5% CO_2: 95% O_2 (v/v) gas mixture
- Krebs-Henseleit Bicarbonate Saline (KHBS): make this solution to contain 118 mM NaCl, 4.7 mM KCl, 2.5 mM $CaCl_2$, 1.2 mM KH_2PO_4, 1.2 mM $MgSO_4$, 25 mM $NaHCO_3$ and 11.1 mM glucose. Equilibrate to pH 7.4 by bubbling 5% CO_2: 95% O_2 (v/v) for about 30 min. Keep the solution oxygenated with a slow stream of the gas mixture.
- Scintillation fluids:
  - Formula 963 (NEN, USA)
  - 0.5% PPO in toluene (w/v)

- Steadie-Riggs Tissue Slicer (Arthur Thomas, USA)   **Equipment**
- McIlwain Tissue Chopper (Brinkman Instruments, USA)
- Sorvall RC-5B Centrifuge SM-24 Rotor and 15-ml Corex tubes
- Water bath shaker
- Scintillation counter and counting vials, Beckman Biovials
- Vortex mixer

1. Sacrifice a sufficient number of male rats by stunning fol-   **Assay for** lowed by decapitation. Quickly remove the cerebrum and   **total IPs** place in ice-cold oxygenated KHBS.

2. Place a Whatman #1 filter circle on the platform of the Steadie-Riggs tissue slicer and wet it with ice-cold KHBS and place the whole apparatus on ice. Transfer each cerebral hemisphere onto the tissue slicer and gently cut two

slices (0.5 mm thick) from each cerebral hemisphere. Place them in ice-cold oxygenated KHBS.

3. Arrange the McIlwain tissue chopper in the cold room with a new blade and set it to 350 μm. Place a disc of Whatman #1 filter paper on the platform of the tissue chopper and wet it with oxygenated KHBS. Carefully spread the cerebral slices flat on the platform and put on the chopper.

   After one cycle, rotate the platform by 90° and put on the chopper for one more cycle. Thus, cross-chopped (350 × 350 μm) slices are obtained. Transfer these to 15 ml of oxygenated KHBS in a tightly stoppered 50-ml Erlenmeyer flask.

4. Gently disperse the slices in the medium with a blunt-ended Pasteur pipette and remove the medium. Add 15 ml of fresh oxygenated KHBS, oxygenate for 1 min and incubate at 37 °C for 15 min in a water bath with gentle shaking.

5. Remove the medium with a Pasteur pipette, add fresh KHBS (15 ml), oxygenate for 1 min, tightly stopper and incubate as in step 4 for 15 min.

6. Repeat step 5, but incubate for 30 min.

7. Remove the medium and transfer 60 μl of gravity-packed tissue slices to each of 12 Beckman Biovials (use a pipetman with a tip, which is cut at the end).

8. To each vial, add 250 μl of oxygenated KHBS containing 5 mM LiCl and 0.3 μM myo-[2-^{3}H]-inositol (The labelled inositol is usually supplied in ethanol. Remove the ethanol and take up the material in KHBS). Gas with 5% CO_2: 95% O_2 (v/v) for 1 min, tightly stopper the vial and incubate at 37 °C for 30 min in a water bath with gentle shaking.

9. After the above incubation, add carbachol in 10 μl such that the final concentration of the agonist is 0.25 to 1 mM.

Again gas the vials for 1 min as in step 8 and incubate at 37 °C for 30 min in a water bath with gentle shaking.

Note. In experiments to test the effect of antagonists, these   **!**
should be added in small volumes (10 µl) at appropriate concentrations 10 min before the addition of the agonist.

10. At the end of incubation, terminate the reaction by adding 940 µl of ice-cold chloroform: methanol (1:2, v/v) and mix the suspension thoroughly on a vortex mixer.

11. Add 0.3 ml of chloroform and 0.3 ml of water to each tube. Let them stand for 30 min and transfer the contents to 15-ml Corex centrifuge tubes. Centrifuge at 1000 g for 5 min in a Sorvall RC-5B with SM-24 rotor to separate the phases.

12. Separate the upper aqueous phase to assay for IPs.

13 Take 0.2 ml of the lower organic phase into scintillation counting vials, evaporate the organic phase, add 10 ml of scintillation fluid and determine radioactivity to assess the incorporation of [^{3}H]-inositol into phospholipids.

14. Dilute 0.75-ml aliquot of the aqueous phase to 3.0 ml with water and apply to a small column of 0.25 g of Dowex AG 1 × 8 resin (200–400 mesh) in the formate form, packed in a Pasteur pipette.

15. Wash the column four times with 3 ml each of 5 mM *myo*-inositol/0.1 M formic acid.

16. Elute the total IPs with 3 ml of 1.0 M ammonium formate/ 0.1 M formic acid into counting vials.

Note. At least three different concentrations of the agonist   **!**
(0.25, 0.5 and 1 mM) should be tested with a control, all in triplicate, which will make a total of 12 vials at step 7.

Calculation: since there may be some variation in the amount of prelabelled tissue taken in each Biovial, the results should be normalised as follows:

Radioactivity in total IPs in a given sample=

$$\frac{\text{dpm in the eluate (step 16) of a given sample}}{\text{dpm in the 0.2 ml organic phase (step 13) of a given sample}}$$

$\times$ Average dpm in the 0.2 ml organic phase (step 13) of all the samples

A two- to three-fold stimulation in the formation of total IPs should be observed when rat cortex is treated with carbachol (1 mM).

2.2.2   Assay for Individual IPs Formed on Agonist Stimulation

The method given here is adapted from Sastry et al. 1992.

Materials
- Male guinea pig (350–400 g)
- Krebs-Henseleit Bicarbonate Saline (KHBS) as described in Section 2.2.1, Materials
- *myo*-[2-³H]-Inositol
- 5% CO_2: 95% O_2 (v/v) gas mixture
- Surfasil (Pierce Chemical, USA)
- Scintillation fluids
  - Formula 963 (NEN-Dupont, USA)
  - 0.5% PPO in toluene (w/v)

Equipment
- Steadie-Riggs tissue slicer, McIlwain Tissue Chopper (see Sect. 2.2.1 for source)
- Water bath shaker
- Scintillation counter
- Beckman Biovials
- Polytron homogeniser
- Sorvall RC-5B centrifuge with SM-24 rotor and 15 ml Corex tubes
- Automated HPLC equipped with Whatman Partisil SAX column (4.6 mm inner diameter ×25 cm), sample applicator, radioactive peak detector and fraction collector. This equipment consists of Waters 600E HPLC pump and controller, a

refrigerated Waters WISP automatic injector, a Radiomatic Beta F10 radioactive flow detector and a Gilson 202 fraction collector.

The counting accuracy of the radioactive flow detector is adequate for peak detection but inadequate for quantitative determination of total radioactivity in a peak. Therefore, each radioactive peak of the flow detector effluent is collected automatically in a single vial in the fraction collector and counted accurately in a scintillation counter. The peaks are recognised by the fraction collector through a peak programme which analyses the flow detector voltage output corresponding to levels of radioactivity.

Note. All glassware should be siliconised by rinsing with an acetone solution of Surfasil and extensive washing with water. !

1. Sacrifice four male guinea pigs by stunning followed by decapitation. Quickly remove the cerebellum, out into two hemispheres and place in ice-cold oxygenated KHBS.
2. Place a petri dish in the inverted position over ice and place each cerebellar hemisphere on it and wet with a drop of oxygenated KHBS. Make approximately 0.5-mm-thick sagittal slices and place them in ice-cold KHBS.

3. Cross-chop ($350 \times 350\,\mu$m) the slices with a McIlwain tissue chopper as described in step 3 in Section 2.2.1, Assay for total IPs.

4. Transfer the cross-chopped slices to 15 ml of oxygenated KHBS in a 50-ml Erlenmeyer flask. Gently disperse the slices with a blunt ended Pasteur pipette and remove the medium. Add 15 ml of fresh oxygenated KHBS, oxygenate for 1 min, stopper tightly and incubate at 37 °C for 15 min in a water bath with gentle shaking.

5. Remove the medium with a Pasteur pipette, add fresh oxygenated KHBS (15 ml), oxygenate for 1 min, stopper tightly and incubate as in step 4 for 15 min.

Assay for individual IPs

6. Repeat step 5 but incubate for 30 min.

7. Remove the medium and preincubate the slices in 4 ml of oxygenated KHBS containing 300 µCi of *myo*-[2-³H]-inositol for 1 h. (If the inositol is supplied in ethanol, ensure that the alcohol is evaporated before taking up the material in KHBS).

8. At the end of the incubation, remove the medium and rinse the slices with 3 × 30 ml of KHBS. This is to remove free *myo*-[2-³H]-inositol.

9. Add fresh oxygenated KHBS (9 ml) to the slices and distribute 0.9 ml of suspension uniformly into 12 Beckman Biovials.

! **Note.** Use a pipetman with a tip cut at the end.

10. Oxygenate the samples for 1 min, tightly stopper and preincubate for 10 min at 37 °C in a water bath shaker.

11. Add 100 µl of 10 mM serotonin (final concentration 1 mM) to make the total volume 1 ml and continue the incubation for 10 min.

12. At the end of the incubation, add 1 ml of 10% TCA to terminate the reaction.

13. Add phytic acid (200 µg) to the quenched samples and homogenise the suspension with a polytron for 20 s.

14. Transfer the homogenates to 15-ml Corex tubes and centrifuge in a Sorvall-RC5B in SM-24 rotor at 7000 rpm for 10 min.

15. Collect the supernatant in separate extraction tubes and wash five times with 5-ml aliquots of diethylether saturated with water (to remove trichloroacetic acid) and neutralise with 0.1 M NaOH to pH 7.0.

16. Take 20-μl aliquots and count radioactivity in a scintillation counter using 10 ml of aqueous counting mixture formula-963 (NEN-Dupont).

17. Suspend the pellets obtained in step 14 in 3.8 ml of methanol:chloroform:water (10:5:3, v/v). Take 50-μl aliquots into scintillation vials, let them dry and determine radioactivity with 10 ml of scintillation fluid (0.5% PPO in toluene) to assess incorporation of labelled inositol into total phosphoinositides.

18. Apply 1.5 ml of the neutralised extract (step 15) to the partisil SAX column and conduct HPLC with the following series of isocratic elutions or linear gradients of ammonium phosphate buffer adjusted to pH 3.8 with H_3PO_4: from 0 to 5 min, a linear gradient from 0 to 10 mM $(NH_4)H_2PO_4$; from 5 to 53 min, an isocratic elution with 10 mM $(NH_4)H_2PO_4$; from 53 to 58 min, a linear gradient from 10 to 160 mM $(NH_4)H_2PO_4$; from 58 to 108 min, an isocratic elution with 160 mM $(NH_4)H_2PO_4$; from 108 to 113 min, a linear gradient from 160 to 600 mM $(NH_4)H_2PO_4$; from 113 to 143 min, a linear gradient from 600 to 800 mM $(NH_4)H_2PO_4$; and from 143 to 188 min, an isocratic elution with 1.75 M $(NH_4)H_2PO_4$. Wash the column with water for 10 min before its next use.

Note. If resolution of IP isomers and IP_2 isomers is not desired, ! a shorter analysis consisting of the following elution programme may be used: from 0 to 30 min, water, from 30 to 35 min, a linear gradient from 0 to 500 mM $(NH_4)H_2PO_4$; from 35 to 65 min, a linear gradient from 500 to 700 mM $(NH_4)H_2PO_4$; and from 65 to 110 min, an isocratic elution at 1.75 M $(NH_4)H_2PO_4$. Wash the column with water for 40 min before its next use.

19. Determine radioactivity in each fraction by counting in a scintillation counter.

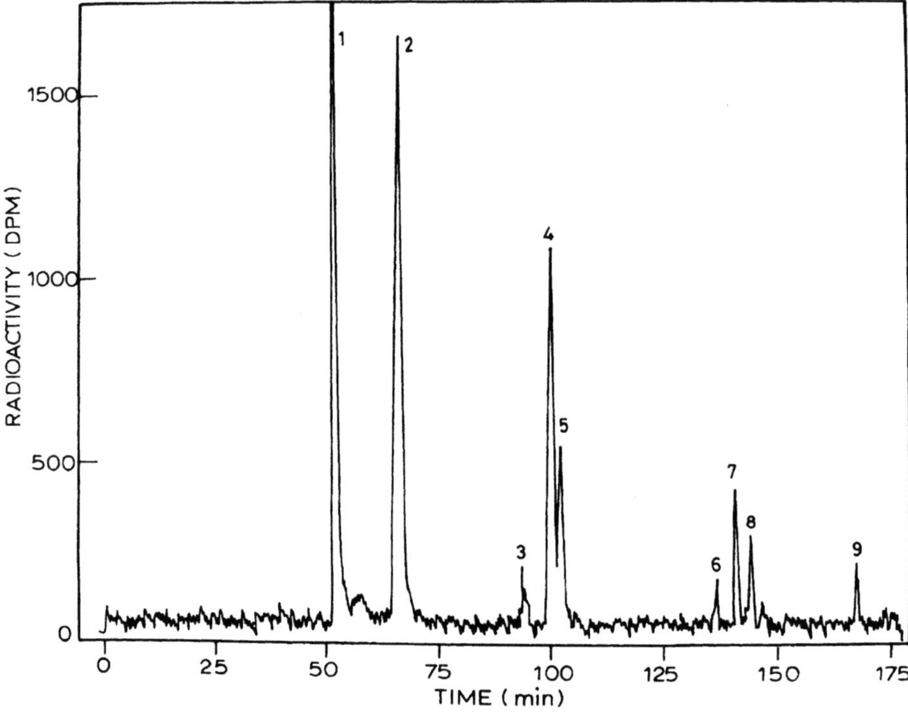

Fig. 5. HPLC separation of [³H]-labelled inositol phosphates. *1* Ins-(1)-P and Ins-(3)-P; *2* Ins-(4)-P; *3* Ins-(1,3)-P$_2$; *4* Ins-(1,4)-P$_2$; *5* Ins-(3,4)-P$_2$; *6* glycerophosphoryl inositol bisphosphate; *7* Ins-(1,3,4)-P$_3$; *8* Ins-(1,4,5)-P$_3$; *9* Ins-(1,3,4,5)-P$_4$. (Sastry et al. 1992)

20. The radioactivity in each fraction of a given sample is normalised to the same amount of tissue as follows:

$$\frac{\text{dpm in a fraction} \times \text{average dpm (in supernatant} + \text{inositol phospholipids) in all samples}}{\text{dpm (in supernatant} + \text{inositol phospholipids) in a given sample}}$$

The various IPs formed in guinea pig cerebellum on stimulation with serotonin and their separation by HPLC is shown in Fig. 5. The experiment should be done with three concentrations of agonist and a control, all in triplicate making a total of 12 samples. The stimulation of various IPs formed by serotonin

(1 mM) varies from 1.3- to 2.0-fold for Ins-(1)-P, Ins-(4)-P, Ins-(3,4)-P_2 and Ins-(1,4,5)-P_3 and from 6- to 8-fold for Ins-(1,3,4)-P_3 and Ins-(1,3,4,5)-P_4.

2.2.3 Receptor Binding Assay for Ins-(1,4,5)-P_3

The first product of hydrolysis of PIP_2 in the PI signal pathway is Ins-(1,4,5)-P_3. Its content can be measured through the use of its specific receptor present in adrenal cortical membranes (Challis et al. 1990).

Materials

- Rats
- KHBS
- Carbachol
- 5% CO_2: 95% O_2 gas mixture, (v/v)
- [^{3}H]-Ins-(1,4,5)-P_3 (Amersham)
- Whatman GF/B Glass fibre filters
- Scintillation fluid: Formula 963 (NEN-DUPONT)

as described in previous sections

Preparation of binding brotein

Take bovine adrenal glands, demedullate and decapsulate to obtain 60–70 g of cortex. Homogenise the tissue in 8 Vol. of ice-cold 20 mM $NaHCO_3$: 1 mM dithiothreitol, pH 8.0 in a blender. Centrifuge the homogenate at 5000 g for 15 min at 4 °C.

Remove the supernatant and homogenise the pellet in 4 vol. of the homogenisation medium and centrifuge as above.

Pool the supernatants and centrifuge at 38 000 g for 20 min at 4 °C. Discard the supernatant and suspend the pellet in the homogenisation medium and recentifuge at 38 000 g for 20 min at 4 °C.

Discard supernatnat and suspend the pellet in the homogenisation medium at a protein concentration of 20 mg/ml, and store at −20 °C in 1-ml batches.

Equipment

- Steadie-Riggs tissue slicer
- Mcllwain tissue chopper
- Water bath shaker
- Scintillation counter
- Millipore filtration set up

as described in previous sections

- Beckman biovials
- Vortex mixer

Receptor binding assay

Steps 1–6. As described under Section 2.2.1, Assay for total IPs.

7. Remove the medium and transfer 50 µl of gravity packed tissue slices to each of 12 Beckman Biovials.

! **Note.** Use a pipetman, with a tip cut at the end.

8. To each vial, add 250 µl of oxygenated KHBS, oxygenate for 1 min, stopper tightly and incubate at 37 °C in a water bath shaker for 60 min.

9. At the end of the incubation, add 10 µl of 31 mM carbachol (final concentration 1 mM) and incubate for 30 s and terminate the reaction by the addition of 300 µl of 10% trichloroacetic acid.

10. Let the acidified samples stand on ice for 15 min, transfer to Corex tubes and centrifuge at 9000 rpm for 10 min in Sorvall RC-5B.

11. Transfer the supernatants to extraction tubes and wash five times with 2 ml of water-saturated diethyl ether. Then add 150 µl of 30 mM EDTA and 150 µl of 60 mM-$NaHCO_3$ to the tissue extract.

12. Carry out the binding at 4 °C in final volume of 120 µl as follows:
    a) Take 30-µl aliquots of the tissue extracts into separate small test tubes in triplicate.
    b) Prepare standards as follows: to 30 µl KHBS, add 300 µl of 10% trichloroacetic acid and extract with 5 × 2 ml of water-saturated diethylether. Then add 150 µl of 30 mM EDTA and 150 µl of 60 mM $NaHCO_3$. Dispense 30-µl aliquots into 15 tubes. To duplicate sets add 0.5, 1.5, 3.0, 4.5, 6.0 and 7.5 pmol of Ins-(1,4,5)-P_3. To the remaining three tubes, add 0.3 mnol of DL-Ins-(1,4,5)-P_3 (to determine non-specific binding).

c) To all the tubes prepared as in (a) and (b), add 30 µl of 100 mM Tris-Cl containing 4 mM EDTA, pH 8.0.

d) Add 30 µl of [^{3}H]-Ins-(1,4,5)-P$_3$ ($\approx$ 7000 dpm/assay) and 30 µl (0.2–0.4 mg of protein) of the adrenal cortical binding protein preparation. Mix the samples intermittently on a vortex mixer for 30 min and keep at 4 °C.

e) Filter rapidly the above incubation mixture through Whatman GF/B glass fibre filters and wash the filter four times with 3 ml each of ice-cold 25 mM Tris-Cl/5 mM NaHCO$_3$/1 mM EDTA, pH 8.0.

f) Transfer the filter discs to scintillation vials, add 15 ml of scintillation fluid (formula 963 Dupont-NEN), let them stand for 12 h and then count for radioactivity.

g) From the radioactivity found in each tube, subtract the non-specific binding (average of triplicates with DL-Ins-(1,4,5)-P$_3$ at step 12b).

h) Plot the specific receptor binding against different concentrations of Ins-(1,4,5)-P$_3$ (from six concentrations at step 12b).

i) From this graph, read the Ins-(1,4,5)-P$_3$ concentration in the sample from the radioactivity observed in it.

j) Express the receptor binding per mg of protein tissue.

k) For protein determination, dissolve the pellets obtained at step 10 in 1 M NaOH and estimate protein by Lowry's method (Lowry et al. 1951).

Note. A kit for the receptor binding assay for Ins-(1,4,5)-P$_3$ is **!** commercially available (Amersham: Biotrak cell signalling assay. TRK100).

3 Phosphatidylcholine Signal Pathway: Assay for Agonist-Stimulated Phospholipase-D Activity

It has already been mentioned that one of the main second messengers produced on receptor activation is diacylglycerol which stimulates protein kinase C. In recent years, evidence was presented showing that diacylglycerol is produced not only from phosphoinositides but also from other membrane phospholipids; in particular, a large number of agonists with a variety of cells were shown to produce diacylglycerol from phosphatidylcholine (PC) which is a major membrane lipid component. In this pathway, PC is hydrolysed by an activated phospholipase D (PLD) and produces phosphatidic acid and choline. Phosphatidic acid is further metabolised by phosphatidic acid phosphatase to produce diacylglycerol and inorganic phosphate (Fig. 6). The activation of this pathway can be assayed by a variety of methods, but the most frequently used method consists of measuring the fomation of phosphatidylethanol on agonist stimulation. The reason for this is that in the presence of 0.1–0.5% ethanol in the medium, PLD catalyses a transphosphatidylation reaction in which the phosphatidyl moiety is transferred to ethanol rather than to water, resulting in the formation of phosphatidyl-ethanol. Give below is a procedure described to assay thrombin-stimulated PLD in human endothelial cells, (Garcia et al. 1992).

Materials
- Human endothelial cells grown in culture to confluence in 35-mm dishes
- M-199 media (GIBCO, Chagrin Falls OH, USA)
- Foetal bovine serum (FBS)
- [^{32}P]-Orthophosphate
- α-Thrombin
- Silica gel H plates containing 1% potassium oxalate
- Phosphatidylethanol (Avanti Biochemicals, Birmingham, USA)
- Scintillation fluid: 0.1% PPO in toluene

Equipment
- Scintillation counter
- Centrifuge
- CO_2-incubator

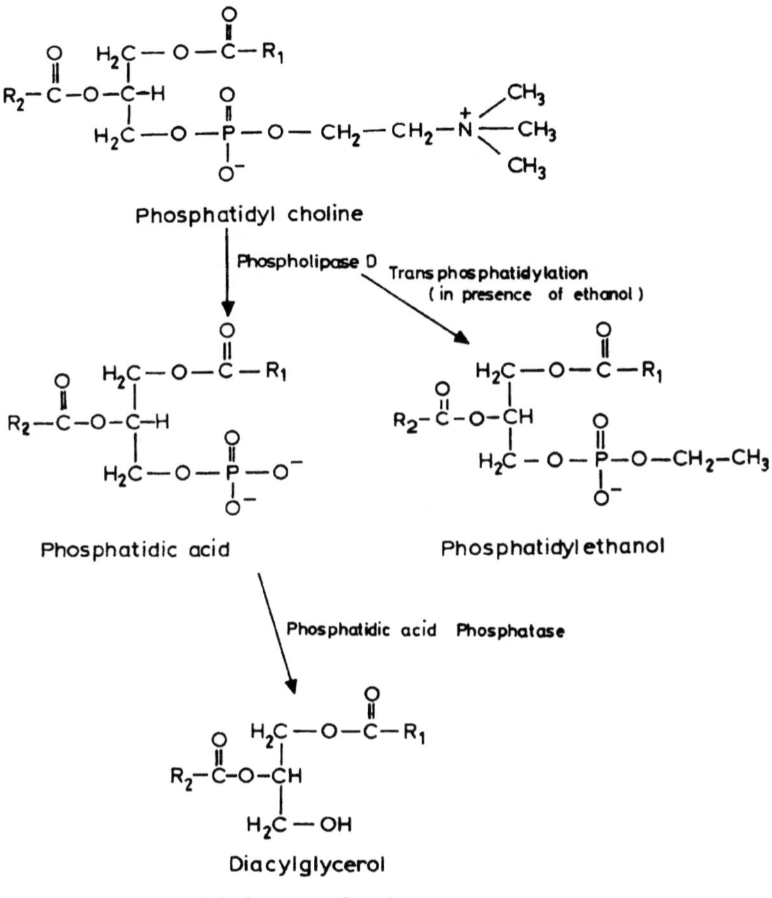

Fig. 6. Phosphatidylcholine signal pathway

Assay for phospholipase D activity

1. Remove the media from the culture dishes where human endothelial cells were grown and add 1 ml of M-199 containing 10% FBS and [^{32}P]-orthophosphate (25 to 30 μCi/dish) and incubate the cells at 37 °C in humidified air containing 5% CO_2 for 18 h.

2. At the end of the incubation period, add α-thrombin to a final concentration of 10 nM (20 picomol in 1 ml containing 4 mM HEPES, 145 mM NaCl, 5.5 mM glucose, 2 mM

CaCl$_2$, 2 mM MgCl$_2$ at pH 7.4 and 0.5% ethanol) and incubate for 15 min.

3. Terminate the incubation by aspirating the medium and by the addition of 1 ml of methanol:HCl (100:1, v/v).

4. Scrap the cells and transfer the suspension to 16 × 125-mm test tubes. Rinse the 35-mm dish with an additional 1 ml of methanol:HCl and transfer it to their respective test tubes.

5. Add 2 ml chloroform and 1.8 ml water. Mix vigorously and centrifuge at 3000 rpm for 15 min.

6. Collect the lower chloroform phase, take it to dryness under N$_2$ and dissolve in about 100 μl of chloroform.

7. Spot the chloroform extract on an activated silica gel H plate containing 1% potassium oxalate. Add 10 μg of phosphatidylethanol and develop the chromatogram using the upper phase of a mixture of ethyl acetate:2,2,4-trimethylpentane:glacial acetic acid:water (65:10:15:50, v/v).

8. At the end of the chromatographic run, remove the plate from the jar, allow it to dry for 10 min and visualise the spots by exposure to iodine vapour and mark the spots.

! Note. In this solvent system, phosphatidylethanol will move with an R$_f$ of 0.3–0.4 and phosphatidic acid will move with an R$_f$ of 0.2. Other phospholipids will remain near the origin.

9. Scrap the phosphatidylethanol spot into scintillation vials and determine radioactivity after the addition of 10 ml of scintillation fluid (0.1% PPO in toluene).

10. Do the experiment (with thrombin), control (without thrombin) and blank (without thrombin and zero min incubation) in triplicates. Subtract the blank values from the experimental and control values.

The radioactivity found in the phosphatidylethanol fraction
on agonist treatment will be four to fivefold higher than in
controls.

4   Sphingolipid Signal Pathway: Assay for Agonist-Stimulated Sphingomyelinase Activity

In the recent past, it was shown that some factors such as
tumour necrosis factor-α (TNF-α), interleukin-1β (IL-1β) and

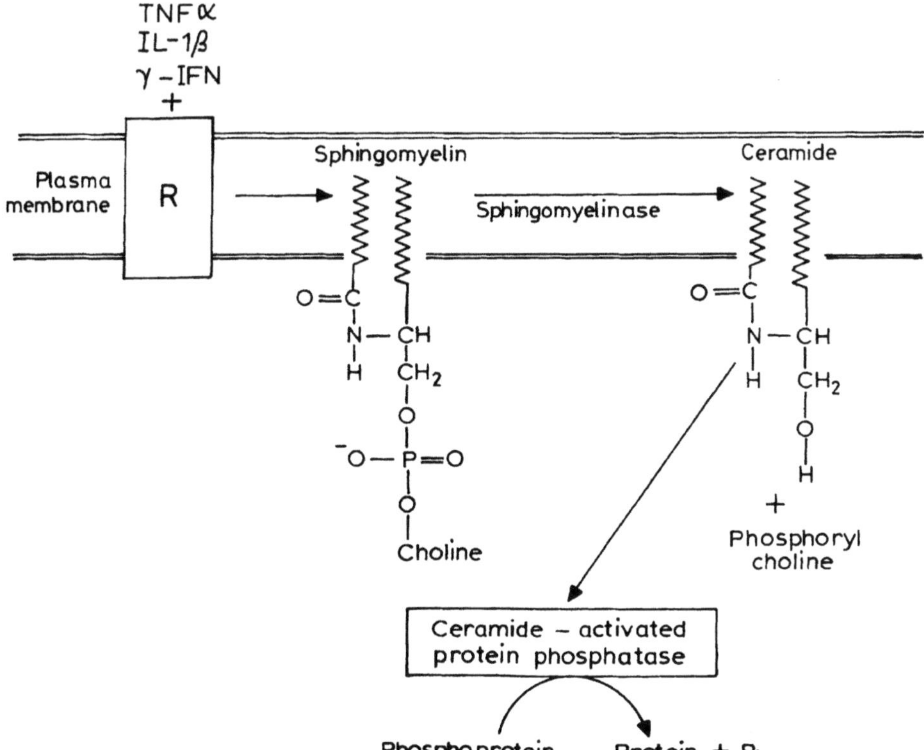

Fig. 7. The sphingomyelin signal transduction pathway

γ-interferon (γ-IFN) employ the 'sphingomyelin pathway' to effect signal transduction by their receptors. In this pathway, agonist binding to the receptor initiates the hydrolysis of sphingomyelin present in the plasma membrane by the activation of a sphingomyelinase, resulting in the formation of ceramide and phosphorylcholine (Fig. 7). Ceramide serves as the second messenger and stimulates a ceramide-activated protein kinase or a ceramide-activated protein phosphatase and thus transduces the cytokine signal. A distinguishing characteristic of this pathway is its participation in antiproliferative mechanisms of cell regulation such as growth inhibition, induction of differentiation and programmed cell death (apoptosis; for reviews, see Hannun and Linardic 1993; Kolesnick and Golde 1994).

Given below is a method to assay the agonist-stimulated hydrolysis of sphingomyelin. This is based on a procedure described by Kim et al. (1991) for experiments with TNF-α and HL-60 cells.

Materials
- HL-60 cells grown in culture to confluence in 35-mm dishes.
- [^{3}H]-choline chloride (80 Ci/nmol) – Dupont – New England Nuclear
- Insulin
- Transferrin
- TNF-α
- Sphingomyelin (Avanti Polar lipids)
- Phosphate-buffered saline (PBS)
- RPMI 1640 medium
- Scintillation fluid (0.5% PPO in toluene w/v)
- Silica gel 60 TLC plates

Assay for sphingomyelinase activity

1. Wash the HL-60 cells grown in culture three times with 2–3 ml of PBS.

2. Add 1 ml of serum-free RPMI 1640 medium containing [^{3}H]-choline chloride (0.5 μCi), insulin (15 μg) and

transferrin (5 µg) and incubate for 48–72 h in a CO_2 incubator at 37 °C.

3. Following the labelling, remove the medium, wash the cells three times with 2–3 ml of PBS, resuspend in 1 ml of serum-free medium and continue incubation for 2 h.

4. Then add 30 pmol of TNF-α in 10 µl (final concentration 30 nM) and incubate for 30 min.

5. Add 2.5 ml of methanol and 1.25 ml of chloroform. Mix well and then add a further 1.25 ml of chloroform and 1.25 ml of water. Mix well and centrifuge at 3000 rpm for 10 min.

6. Take 0.25 ml of the lower chloroform phase into scintillation vials, evaporate the solvent and determine the radioactivity with 10 ml of scintillation fluid. This radioactivity ×10 will give the total radioactivity in the lipid extract.

7. Take 1.5 ml of the chloroform phase into a conical centrifuge tube, evaporate the solvent, take it up in 50 µl of chloroform, add 10 µg of sphingomyelin as carrier and spot on a silica gel 60 TLC plates and develop the chromatogram with the solvent system chloroform:methanol:acetic acid:water (60:30:8:5, v/v). Let the plate dry for 10 min and visualise the spots with iodine.

8. Scrap the sphingomyelin spot into scintillation vials and determine radioactivity after the addition of 10 ml scintillation fluid.

Note. Do the experiment (with TNF-α), control (without TNF-α) and blank (without TNF-α, zero min incubation) in triplicates.

Normalise the radioactivity found in the sphingomyelin spot as follows:

Radioactivity in sphingomyelin in a given sample=

$$\frac{\left(\begin{array}{c}1.66^* \times \text{dpm in sphingomyelin spot} \\ \times \text{ dpm in total lipid extract}\end{array}\right)}{(\text{average dpm in total extract of all the samples (step)})}$$

*This factor is to account for the 2.5 ml of chloroform extract of which only 1.5 ml is taken for TLC.

A reduction of about 40% in the label in sphingomyelin was observed when HL-60 cells were treated with TNF-α (Kim et al. 1991).

5   Glycosylphosphatidylinositol Signal Pathway: Assay for the Hydrolysis of Glycosylphosphatidylinostol on Agonist Stimulation

A novel second messenger system has been implicated in insulin, nerve growth factor (Chan et al. 1989) and interleukin-2 signalling (Eardley and Koshland 1991). With these ligands, a set of structurally related glycosylphosphatidyl inositol (Gly-PI) molecules are believed to be involved (Gaulton et al. 1988). These molecules reside in the membrane and contain a distinct hydrophobic domin usually 1,2-dimyristoylacylglycerol and a PI that is glycosidically linked to a glycan moiety through glucosamine. When the agonist binds to a receptor, a specific phospholipase C is activated and Gly-PI is hydrolysed, resulting in the formation of dimyristoyl glycerol in the membrane and release of inositol phosphate-glycan into the cytoplasm (Fig. 8). The second messenger function of dimyristoylglycerol has not been established, but evidence has been presented suggesting a second messenger role for inositol phosphate-glycan. This system has been implicated in Interleukin-2 signal transduction in lymphoma cells (Eardley and Koshland 1991). Based on this report, the following procedure is described.

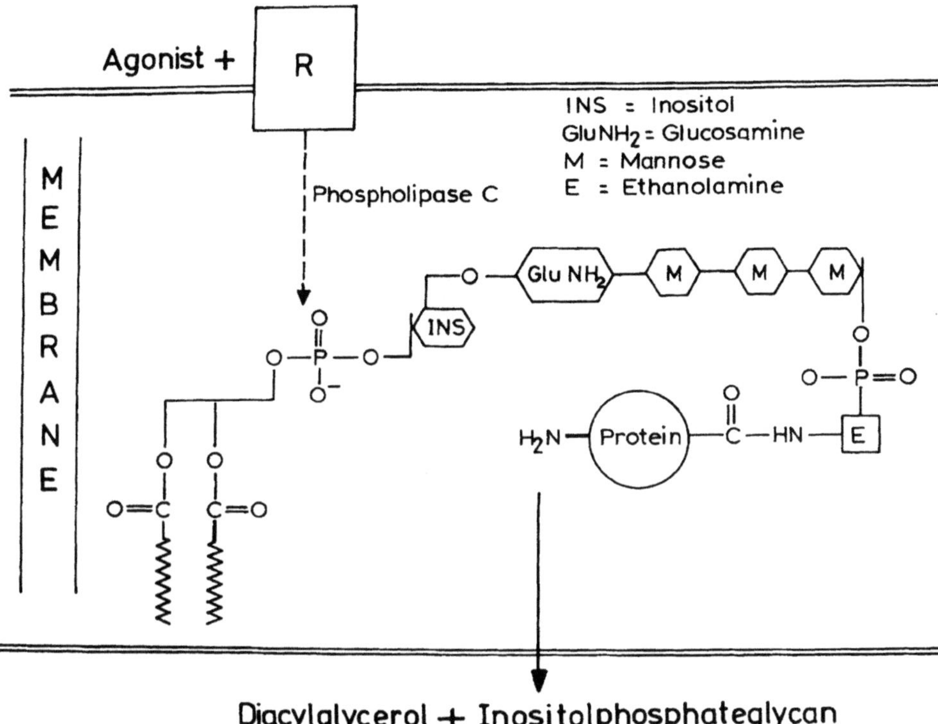

Diacylglycerol + Inositolphosphateglycan

Fig. 8. Glycosylphosphatidylinositol signal pathway

- BCL1, lymphoma cell line (IL-2 inducible B cell line)   **Materials**
- Interleukin-2 (recombinant human IL-2) – (Amgen, USA)
- [^{3}H]-myristic acid (Amersham)
- [^{3}H]-inositol (Amersham)
- Dipalmitoyldiglyceride (Sigma)
- RPMI 1640 Medium
- Hank's balanced salt solution
- Foetal bovine serum

- Silica gel G plates (Analtech)   **Equipment**
- Chromatography jars
- Centrifuge
- HPLC with an analytical Partisil 10 SAX column (Whatman)

Assay with [³H]-myristic acid

1. Culture BCL1 cells (5×10^5 cells/ml) in RPMl 1640 medium containing 0.5% foetal bovine serum (low serum pro-liferative condition) overnight with [³H]-myristic acid ($2\,\mu Ci$).

2. Wash the cells twice with Hank's balanced salt solution.

3. Take 10^6 cells in 0.8 ml in test tubes (in triplicate for control and Interleukin-2 treatment).

4. Add IL-2 (150 pM final concentration) in $10\,\mu l$ and incubate for 3 min.

5. Stop the reaction by the addition of 3 ml of chloroform: methanol: 1N HCl (100:200:1, v/v) followed by 1 ml of chloroform and 1 ml of 50 mM formic acid. Mix thoroughly and centrifuge at $500\,g$ for 10 min.

6. Take 1 ml of the chloroform phase and add $100\,\mu g$ of cold dipalmitoylglycerol and evaporate the solvent.

7. Take it up in about 20–$30\,\mu l$ of chloroform and quantitatively spot on a silica gel G TLC plate and develop with n-hexane:diethyl ether:acetic acid (65:35:1, v/v).

8. Let the TLC plate dry for 10 min. Expose it to iodine vapour and mark the diacylglycerol spots.

9. Scrap them into scintillation vials, add 10 ml scintillation fluid (0.5% PPO in toluene) and determine radioactivity.

In samples treated with IL-2, three- to fourfold higher radioactivity will be observed, indicating stimulated formation of dimyristoyldiglycerol.

Assay with [³H]-inositol

1. Culture BCL1 cells (10^6 cells/ml) RPMI 1640 medium containing 0.5% foetal bovine serum for 2 h with [³H]-inositol ($3\,\mu Ci$).

2. Wash the cells twice with Hank's balanced salt solution.

3. Take 10^6 cells in 0.8 ml in test tubes (in triplicate for control and IL-2 treatment).

4. Add IL-2 in $10\,\mu l$ ($750\,pM$ final concentration) and incubate for 2 min.

5. Stop the reaction by the addition of 3 ml of chloroform:methanol:1N HCl (100:200:1, v/v) followed by 1.0 ml of chloroform and 1.0 ml of 50 mM formic acid. Mix thoroughly and centrifuge at $500\,g$ for 10 min.

6. Take 1.5 ml of aqueous phase, evaporate to remove methanol and pass through a C_{18} Sep-Pak column to remove residual lipids. Lyophilise the sample and dissolve in a small amount of water ($20\,\mu l$). Subject it to HPLC on an analytical Partisil 10 SAX column as follows: 0–5 min isocratic elution with 20 mM trimethylamine formate (pH 4.5); 5–30 min linear gradient elution with 20 mM-1 mM trimethylamine formate (pH 4.5) at 1 ml/min.

7. Collect 1-ml fractions and determine radioactivity.

8. Under these conditions, inositol phosphate-glycan elutes in about 17 min.

9. Severalfold increase in radioactivity in inositol phosphate-glycan peak in the cells treated with IL-2, as compared to controls will be observed.

6   Concluding Remarks

There is now overwhelming evidence for the participation of membrane lipids in cellular recognition and signal transduction phenomena. At present, it is well established that polyphosphoinositides, phosphatidylcholine, some sphingolipids and the membrane anchor components, i.e. glycosylphosphatidylinositols, generate second messengers on receptor activation. It is clear that phospholipase C and phospholipase D are involved in signal transduction but evidence for the involvement of phospholipase A_2 is not yet adequate. However, this is now a very active field of investigation and it is to be expected that a role for many other membrane lipids in signal transduction will soon be found.

In this chapter, detailed procedures for studying this phenomena have been described by taking examples of some tissues and cells in culture with a particular agonist. These may be taken as illustrative examples and the procedures can be adopted for other tissues and agonists.

References

Blackstone CD, Supattapone S, Snyder SH (1989) Inositolphospholipid-linked glutamate receptors mediate cerebellar parallel-fibre-Purkinje cell synaptic transmission. Proc Natl Acad Sci USA 86:4316–4320

Challis RAJ, Chilvers ER, Willcocks AL, Nahorski SR (1990) Heterogeneity of [^{3}H]-inositol-1,4,5-triphosphate binding sites in adrenal-cortical membranes: characterisation and validation of a radioreceptor assay. Biochem J 265:421–427

Chan BL, Chao MV, Saltiel A-R (1989) Nerve growth factor stimulates the hydrolysis of glycosyl phosphatidylinositol in PC-12 cells: a mechanism of protein kinase C regulation. Proc Natl Acad Sci USA 86:1756–1760

Chandrasekhar M, Hokin LE (1986) The role of phosphoinositides in signal transduction. J Memb Biol 89:193–210

Dawson RMC, Eichberg J (1965) Diphosphoinositide and triphosphoinositide in animal tissues: extraction, estimation and changes post mortem. Biochem J 96:634–643

Eardley DD, Koshland ML (1991) Glycosylphosphatidylinositol: a candidate system for interleukin-2 signal transduction. Science 251:78–81

Garcia JGN, Fenton JW, Natarajan V (1992) Thrombin stimulation of human endothelial cell phospholipase D activity regulation by phospholipase C, protein kinase C and cyclic adenosine monophosphate. Blood 79:1–7

Gaulton NG, Keely KL, Powlowski J, Mato JM, Jarett L (1988) Regulation and function of an insulin-sensitive glycosyl-phosphatidylinositol during T-lymphocyte activation. Cell 53:963–970

Hannun YA, Linaridic CM (1993) Sphingolipid breakdown products: antiproliferative and tumour-suppressor lipids. Biochim Biophys Acta 1154:223–236

Kim MY, Linardic C, Obeid L, Hannun Y (1991) Identification of sphingomyelin turnover as an effector mechanism for the action of tumour necrosis factor-α and γ-interferon. J Biol Chem 266:484–489

Kolesnick R, Golde DW (1994) The sphingomyelin pathway in tumour necrosis factor and Interleukin-1 signalling. Cell 77:325–328

Lagos N, Vergara J (1990) Phosphoinositides in frog skeletal muscle: a quantitative analysis. Biochim Biophys Acta 1043:235–244

Lowry OH, Rosenbrough NJ, Farr AL, Randall RJ (1951) Protein measurement with the Folin phenol reagent. J Biol Chem 193:265–275

Sastry PS, Dixon JF, Hokin LE (1992) Agonist-stimulated inositol polyphosphate formation in cerebellum. J Neurochem 58:1079–1086

Chapter IX Liposomes: Preparation and Membrane Protein Reconstitution

Maïté Paternostre[1], Michel Ollivon[1] and
Jacques Bolard[2]

1 Background

1.1 Definition of Liposomes

Liposomes are hollow microspheres, their membrane is composed of one or several lipid bilayers and encapsulates a small volume of the medium in which they have been prepared. This morphology enables them to encapsulate water-soluble compounds in their internal volume and liposoluble ones in the bilayers. Since they are weakly water-soluble, amphiphilic compounds may also be encapsulated into liposomes. Depending on their composition, these aggregates may be chemically stable for durations from hours to years. Physical stability, which depends on liposome concentration and particle interactions, may also be achieved for long periods on specific bilayer stabilisation.

The ability of liposomes to reproduce the organisation and the main functions of cells has made them a very interesting model for the study of molecular mechanisms existing in living systems. However, designing well-defined liposomes suitable for such studies is very delicate, and requires precise prepara-

[1] Equipe Physicochimie des Systemes Polyphases, URA CNRS 1218, Universite Paris Sud, 5 rue Jean-Baptiste Clement, 92296 Chatenay-Malabry, France

[2] Laboratoire de Physique et Chimie Biomoleculaires, URA CNRS 198, Universite Pierre et Marie Curie, 4 Place Jussieu, 75252 Paris Cedex 0.5, France

tion procedures and control techniques. This chapter makes it possible for the scientist who is inexperienced with liposomes to prepare and confidently use the standard methods of liposome preparation. In order to provide basic tools for the standard characterisation of the preparations, some of the techniques used for liposome control are also described.

Numerous types of liposomes exist: the size, shape (which is not necessarily that of a sphere), number of lamellae, the lipid composition of liposomes, as well as the production techniques, may be different and should be adapted according to the final use of these vesicles.

1.2   Use of Liposomes

The interest in liposomes is growing in fields as different as physics, physical chemistry, biophysics, pharmacology, cosmetology or food research. For physicists and physical chemists, liposomes represent a system in which forces between amphiphiles can be studied. For biophysicists and biologists, the liposomes are the best membrane model in which membrane proteins can be inserted and studied. For pharmacologists, liposomes are potential drug carriers.

Although liposomes are potentially good candidates for drug delivery and targeting, the clinical use of liposomes introduces a number of technical problems, such as sterility and conservation of the products, which specialists in this field can resolve, the liposomes described below are intended for laboratory use.

The size (small, large or giant liposomes) and the structure (uni- or multi-lamellar) of the liposomes are determined by the technique used for their formation, whereas the properties of the vesicle membranes (fluidity, permeability, chemical stability) are given by the type of lipid used for membrane composition. On the other hand, the physical stability (aggregation and fusion, swelling and bursting) of the liposomes is influenced both by the size of the liposomes and by the nature of the lipid used for the membrane composition. Liposomes can be used either for the encapsulation of hydrophilic molecules in the internal aqueous medium or for the incorporation of a hydrophobic and/or amphiphilic molecule in the membrane. The ability of the liposomes to encapsulate a high concentration of hydrophilic molecules or to incorporate a high amount of hydrophobic ones depends on the technique used. Therefore, it is important to choose appropriate lipids and techniques, depending on the use of the liposomes. In this chapter, different types of procedures of liposome formation and the resulting type of liposomes formed are discussed. Guidelines concerning the choice of both lipids and preparations, according to the required application, are also given. There are four different techniques, leading to the formation of vesicles of different sizes (Fig. 1): small unilamellar vesicles (SUV), large unilamellar vesicles (LUV) and very large unilamellar vesicles (VLUV).

The last part of this chapter deals with the problems of membrane protein reconstitution. Since the most powerful technique to incorporate membrane proteins is to use detergents, the formation of lipidic vesicles using detergent is discussed on one hand and a strategy to determine the optimal reconstitution conditions is provided on the other.

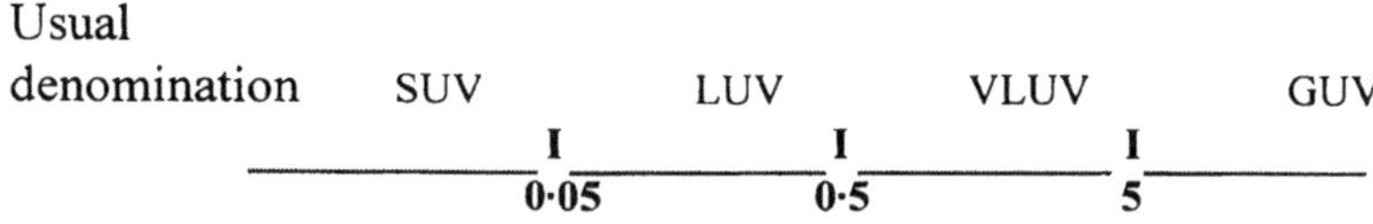

Fig. 1. Liposomal vesicles

2 Overview of the Main Techniques of Preparation

Liposomes are prepared from lamellar phase-forming lipids such as phospholipids. These may be mixed to other lipids in order to come closer to membrane properties (see Sect. 3). The substances to be encapsulated into liposomes should be introduced during their process of formation at two different stages, depending on their nature: lipophilic substances are introduced as soon as the lipids are mixed, while hydrophilic substances are added at the step preceding liposome closure. Lipids are mixed using either solvent or surfactant solubilization. In both cases, the step of either solvent or surfactant removal leads to the formation of liposomes (Fig. 2).

Note. Irrespective of the method of preparation complete **!** solubilisation of lipid (i.e. formation of an isotropic solution) is necessary.

Many techniques of liposome formation use mechanical **Mechanical** dispersion of lipids (Table 1). The structure (bilayer or multilayer) and the size (μm or nm range) of the liposomes formed are closely related to the strength of the mechanical shear constraint to which initial suspensions are submitted (Table 1).

One general disadvantage of all the techniques using organic **Organic** solvent is the denaturation of labile molecules, which occurs **Solvents** when in contact with the solvent. Moreover, attention must be paid to the temperatures of phase transition of the lipids (Tc) and to the boiling of the solvent. The liposomes must be prepared at a temperature higher than the Tc of the lipid used and, as a consequence, a solvent with a boiling temperature higher than the Tc has to be chosen (Table 2).

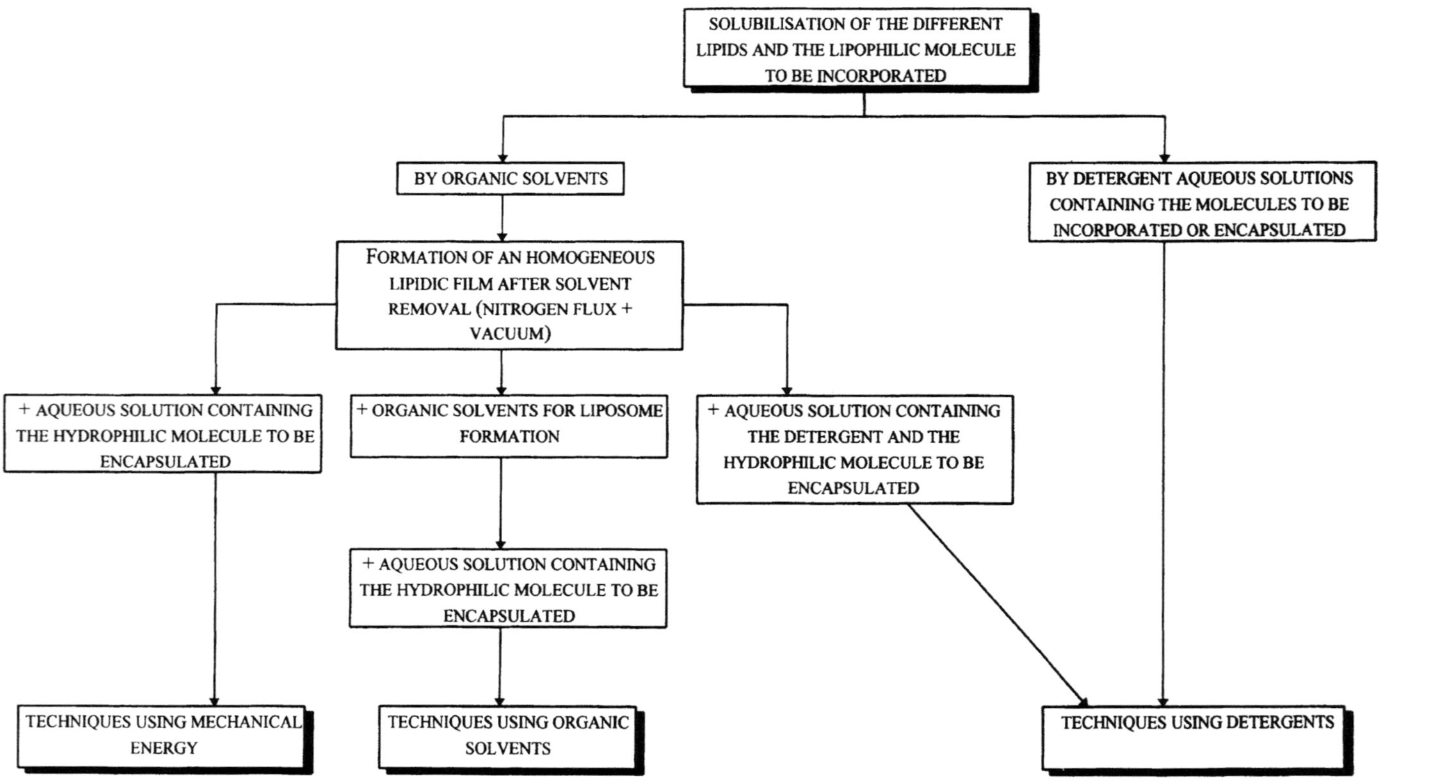

Fig. 2. General procedure of liposome formation

Table 1. Mechanical processes

Process, reference	Energy	Production	Structure	Size	Size dispersity	Internal volume	% Encapsulation	Comments
Hydration Bangham et al. (1965)	Mechanical stirring	1 ml to 1 l	MLV	D > 1 μm	Very high size polydispersity	0.5–4 μl/μmol of lipids	2–15	-The encapsulation strongly depends on stirring
French press Barenholz et al. (1979) Hamilton and Guo (1984) Lelkes (1984)	Pressure + Extrusion	Laboratory to industrial use	SUV	30–50 nm	Small polydispersity	0.2–1.5 μl/μmol of lipids	5–25	-The % of encapsulation strongly depends on the lipid concentration -Increase of temperature at each pass in the French press
Extrusion through polycarbonate membranes Hope et al. (1985, 1986) Olson et al. (1979)	Pressure + pore calibrated membrane	Laboratory use	LUV	About 100 nm	Very small polydispersity for the small size (100 nm)	3 μl/μmol of lipids	5–30	-Apparatus or small syringes adapted on the filters -Calibration of the vesicles
Microfluidiser Mayhew et al. (1984)	High pressure (10 000 psi)/ →collision of vesicles (high speed)	Laboratory use	LUV	100–200 nm			Up to 70	-Possibility of very high lipid concentration

Table 1. *Continued*

Process, reference	Energy	Production	Structure	Size	Size dispersity	Internal volume	% Encapsulation	Comments
Dispersion of lyophilised liposomes Shew and Deamer (1985) Kirby and Gregoriadis (1984) Oshawa et al. (1984)		Quasi-industrial use	MLV	About 1 μm	High polydispersity		Up to 40	-Rehydration of preformed lyophilised SUV with small amount of a solution containing the molecule to be encapulated -High amount of encapsulation
Freeze-thawing Pick (1981) MacDonald and MacDonald (1983) Kashara and Hinkle (1977)			LUV MLV		High polydispersity	10 μl/μmol of lipids 50 μl/μmol of lipids	30 88	-Freezing of preformed SUV or MLV followed by thawing at room temperature -Requires the presence of negatively charged lipids
Sonication Saunders et al. (1962) Huang (1969) Papahadjopoulos and Miller (1967)	Ultrasound irradiation	Laboratory use	SUV	20–50 nm	Small polydispersity	0.2–1.5 μl/μmol of lipids	0.1–1	-Denaturation of labile molecules -Very low rate of encapsulation -Instability of the vesicles

Table 2. Techniques using organic solvents

Techniques, reference	Production	Structure	Size	Size dispersity	Internal volume	% Encapsulation	Comments
Injection of ethanol Kremer et al. (1977) Batzri and Kom (1973)	Injection of a lipid-containing ethanol solution into an aqueous solution	SUV-LUV	30–150 nm				-The size and the rate of encapsulation depend on the rate of ethanol solution injection into the aqueous phase -Dialysis of ethanol required after liposomes formation
Reverse phase evaporation Szoka and Papahadjopoulos (1978) Szoka et al. (1980)	-Mixing of lipids in an organic solvent non-miscible with water -Slow evaporation -Transition between inverse micelles and liposomes	MLV LUV	500 mm 150 nm after extrusion through polycarbonate membranes	Very high polydispersity Very small polydispersity		65 (10 mM NaCl) 20 (500 mM NaCl) 12	-Depending on the use, formation of MLV with very high rate of encapsulation or of stable and monodisperse LUV with low rate of encapsulation
Ether infusion Deamer and Bangham (1976) Hiezen et al. (1978)	-Injection of 2 ml solvent solution containing the lipids into 4 ml aqueous solution (55–65 °C, low pressure) at a rate of 0.2 ml/min -Ethyl ether or ethyl ether/methanol -Petroleum ether	LUV	150–250 nm		13–21 µl/µmol of lipids	2 36–46	-The low rate of encapsulation is due to the low solubility of phospholipids in ethyl ether -Increase of the encapsulation rate is achieved by using petroleum ether but only few lipids are soluble in this solvent

Detergents These techniques are mainly based on the properties of detergents on one hand and on the properties of mixed amphiphilic systems on the other. A "detergent" is an amphiphilic molecule which is soluble in its monomeric form until it reaches its critical micellar concentration (CMC). Above CMC, it aggregates to form micelles. When lipids are added to a solution of micelles, they generally incorporate the micelles and form mixed lipid-detergent micelles. The elimination of the detergent from these mixed lipid-detergent micelles leads to the spontaneous formation of generally unilamellar liposomes, whose size and properties depend not only on the nature of both the lipids as well as the detergent, but also on the elimination rate of the detergent, the temperature, the ionic strength etc. Different methods of liposome formation using detergent are essentially based on techniques which allow specific detergent elimination. In general, detergents with a high CMC are more easily eliminated than those with a lower one because of their partition between solution and vesicles (Table 3).

3 Choice of Lipids and of Preparation Techniques as a Function of the Application

3.1 Choice of Lipids as a Function of the Application

The physicochemical characteristics of the vesicle membranes are essentially governed by their lipidic composition. All membrane lipids are amphipathic in nature. This character is largely responsible for supramolecular bilayer structure of membrane, i.e. the polar head groups of molecules are oriented towards aqueous medium, whereas the hydrophobic cores of molecules are protected from water and inserted into the bilayer.

The lipids are divided into two different classes as a function of their structures:

Table 3. Techniques using detergents

Techniques	CMC range	Examples of detergent	Reference	Comments
Dialysis	CMC > 1 mM	Sodium cholate Taurocholate Sodium deoxycholate Octyl glucoside	Kagawa and Racker (1971) Milsman et al. (1978) Mimms et al. (1981) Rhoden and Golden (1979) Schwendener et al. (1981)	This technique can only be employed for high CMC detergent The elimination is long This technique has been extensively employed for elimination of ionic detergent and in particular biliary salts
Gel exclusion chromatography	CMC > 1 mM	Sodium cholate Taurocholate Sodium deoxycholate Octyl glucoside	Enoch and Strittmatter (1979) Bruner et al. (1976)	This technique can only be employed for high CMC detergent
Dilution	CMC > 1 mM	Octyl glucoside	Ollivon et al. (1988)	The dilution technique induces a decrease in detergent to lipid ratio in the structure in favour of quantity of monomeric detergent in solution Only valid for high CMC detergent This technique induces an important dilution of the sample As the detergent is not eliminated, this procedure has to be followed by dialysis or gel exclusion chromatography This technique allows to monitor the rate of detergent elimination
Contact with polystrene beads	CMC < 1 mM CMC > 1 mM	TX-100 C12E8 OG	Horigome and Sugano (1983) Moriyama et al. (1984) Phillipot et al. (1983) Phillipot et al. (1985) Rigaud et al. (1988)	This technique can be used for non-ionic detergent This technique is very efficient for the elimination of low CMC detergent Very large unilamellar vesicles can be prepared

- Lipids which contain fatty acids (glycerolipids, phospholipids and sphingolipids).
- Lipids which do not contain fatty acids, the most representative molecule of this group being cholesterol.

Lipids belonging to each class are briefly discussed in Chapter I, this Volume.

The most common lipids used for model membrane reconstitution are either a mixture of phospholipids (mixture of phosphatidylcholine and phosphatidic acid, phosphatidylcholine and phosphatidylserine) or phospholipids with other lipids belonging to the same class (sphingolipids or glycolipids) or to another class (for example, cholesterol). In order to confer unique properties to the membrane, different lipid mixtures are used. In this chapter, information about the choice of the lipid as a function of the type of the membrane desired is given. In particular, the influence of membrane composition on the fluidity and the permeability of the membrane, and also on the chemical and physical stability of vesicles is also reviewed.

In the use of natural phospholipids (e.g. egg phosphatidylcholine), or synthetic phospholipids containing polyunsaturated fatty acids, two possibilities present themselves by which lipids can be degraded, e.g. by hydrolysis or by oxidation. The latter is easily reduced by the use of highly purified materials, the addition of an antioxidant (α-tocopherol or butyl hydroxytoluene, BHT) or the preparation of the liposomes under oxygen free atmosphere (nitrogen or argon). The degradation of liposomes by hydrolysis can be prevented only by lyophilisation; however, its use is not always possible because the rehydration of the liposomes very often leads to the leakage of the encapsulated molecules and to drastic changes in their shape and size. In order to reduce hydrolytic degradation, storage at low temperature (4–6 °C), after adjustment of the buffer solution at an approximately neutral pH, is preferred. In cases where buffer species tend to accelerate the hydrolysis, it is better to use a lower concentration of buffer allowing a stable pH (for a review see Grift and Crommelin 1993).

The fluidity of the membrane, or, more precisely, the lateral and rotational mobility of the molecules, depends on the acyl chain region, i.e. the hydrophobic part of most of lipids. A bilayer of a pure lipid (synthetic lipid) will undergo a gel to liquid-crystalline phase transition at a defined temperature, which depends solely on acyl chain length, degree of unsaturation and type of polar group (Table 4), whereas most natural membranes are composed of lipids, with a broad distribution in length, number and position of unsaturations of the acyl chains, and also have different polar groups. As a result, natural membranes are frequently in the liquid-crystalline phase at room temperature and even below 0 °C. The first implication of this behaviour is that, depending on the lipid source, i.e. synthetic or natural, obtaining of membranes in a gel or liquid-crystalline phase is possible. This membrane property has a direct influence on the lateral mobility of the protein inserted in the bilayer and also on membrane permeability.

The polar head group of the lipids may be non-ionic or zwitterionic (phosphatidyl-choline, -ethanolamine etc., Table 4), presenting no net charge or, on the contrary, ionic (phosphatidylserine, phosphatidic acid), the ionisation depending in particular on the pH of the aqueous medium. In general, presence of a net charge at the surface of the membrane reduces the aggregation of vesicles by electrostatic repulsion and, consequently, the fusion process, resulting in a better physical stability of liposomes. In most cases, a small percent of ionic lipids (5 to 10%) is sufficient to reduce these phenomena (Lesieur et al. 1991, 1993). For example, addition of 10% egg phosphatidic acid has been extensively used to stabilise large unilamellar vesicles made of phosphatidylcholine extracted from egg yolk (Paternostre et al. 1988).

Table 4. Lipids used for liposome preparation and their physicochemical characterisics

	Abbreviation	Origin	Net charge at pH 7.4	T °C[a]	MW
Natural lipids					
Phosphatidylcholine	EPC	Egg yolk	0	−15 to −7	763
Phosphatidylcholine	SPC	Soyabean	0	−	780
Sphingomyelin	SPM	Bovine brain	0	32	730
Phosphatidylserine	PS	Bovine brain	−1	6 to 8	800
Phosphatidylethanolamine	PE	Bovine brain	0	−	743
Phosphatidic acid	PA	Egg yolk	−2	−	696
Cholesterol	Chol	Porcine liver	0	−	387
Synthetic lipids					
Dilauroyl PC	DLPC		0	−1.8	640
Dimyristoyl PC	DMPC		0	23	696
Dipalmitoyl PC	DPPC		0	41.5	752
Distearoyl PC	DSPC		0	58	808
Dioleoyl PC	DOPC		0	−22	804
Dimyristoyl PE	DMPE		0	40	635
Dipalmitoyl PE	DPPE		0	60	692
Distearoyl PE	DSPE		0	74	748
Dioleoyl PE	DOPE		0	−16	744
Dimyristoyl PG	DMPG		−1	23	666
Dipalmitoyl PG	DPPG		−1	41.5	722
Distearoyl PG	DSPG		−1	55	779
Dioleoyl PG	DOPG		−1	−18	775
Dipalmitoyl PS	DPPS		−1	51	735
Dicetylphosphate	DCP		−1		546
Hexadecyldiglycerol	C16G2		0		390
Stearylamine	SA		+1		269

[a] These temperatures were determined for lamellar phases. They actually vary with the type of vesicles, in a range from 2 to 3 °C.

Literature regarding the effect of cholesterol on lipid membranes is abundant and explains the complexity of its effects. The first observations on the interaction of cholesterol with phospholipids were obtained on monolayers. It was established that the area per molecule of lecithin plus cholesterol was considerably smaller than its area measured on a monolayer constituted by pure lecithin (Lecuyer and Dervichian 1969; for review see Phillips 1972, and Papahadjopoulos and Kimelberg 1974). This so-called condensation effect of cholesterol was also observed on phospholipid bilayers (Ipsen et al. 1990). It appears that the addition of cholesterol produces a more condensed state, where the motion of hydrocarbon chains of the phospholipids is restricted. Phospholipids are also prevented from crystallisation in the hexagonal packing characteristics of the pure phospholipids below their liquid-crystalline transition temperature. Thus, it can be stated that:

- Addition of cholesterol induces the disappearance of the transition between gel to liquid-crystalline phases.
- Mixed cholesterol/phospholipid membrane is rendered more impermeable to water, electrolytes (such as anions and cations) and non-ionic molecules, compared to the pure phospholipid membranes (Blok et al. 1976; Fettiplace and Haydon 1980; Cullis and Hope 1985).

As a consequence, liposomes containing cholesterol, exposed to osmotic gradient, may collapse because they also loose their capacity to swell.

On the other hand, additives such as cholesterol, which are not able to form a bilayer by themselves, will participate in its formation when associated with other lipid compounds without this capability, for example, the stoichiometric association of cholesterol and hexadecyl-diglycerol ether (C16G2), which is a lipid containing a single saturated aliphatic chain with a diglycerol group. This last group, which occupies a larger area per chain than EPC, prevents this compound from forming bilayers; for example, when associated with cholesterol, it leads to the formation of a mixed lipid bilayer (Vanlerberghe et al. 1978). In this particular case, cholesterol stabilises a bilayer structure and allows formation of very stable and impermeable vesicles.

In addition, attention must be paid to the composition of the buffer solution used for liposome formation, in particular when charged lipids like phosphatidic acid or phosphatidylserine are introduced into the liposome composition. Specifically, the buffer must be free of calcium. This divalent cation has two distinct effects on liposomes:

- Aggregation and fusion of liposomes (Papahadjopoulos et al. 1977)
- Lateral phase separation of lipids (Welti and Glaser 1994).

Note. In order to be sure that the buffer is free of Ca^{2+}, deionised water is used for the preparation of buffer and 1 mM of EDTA is also added to the solution. Attention must also be paid to the choice of the lipid salt used, i.e. choosing the sodium salt of PA instead of the Ca^{2+} one.

3.2 Choice of Preparation Techniques as a Function of the Application

The nature of the molecules to be encapsulated dictates the choice of the type of liposome, e.g. encapsulation of hydrophilic molecules depends on internal aqueous compartment volume, while the lipidic bilayer volume drives the incorporation of hydrophobic or amphipathic ones. The potential capacities of the liposomes for encapsulation essentially depend on their size, internal volume and lipid concentrations. Figure 3 illustrates the variations of these parameters as a function of vesicle size.

It is evident from Fig. 3 that large liposomes (LUV) are more adapted than small ones (SUV) for encapsulation of hydrophilic molecules. Since their encapsulation ratio depends on the lipid concentration, the processes allowing both the formation of very large unilamellar vesicles and the use of very

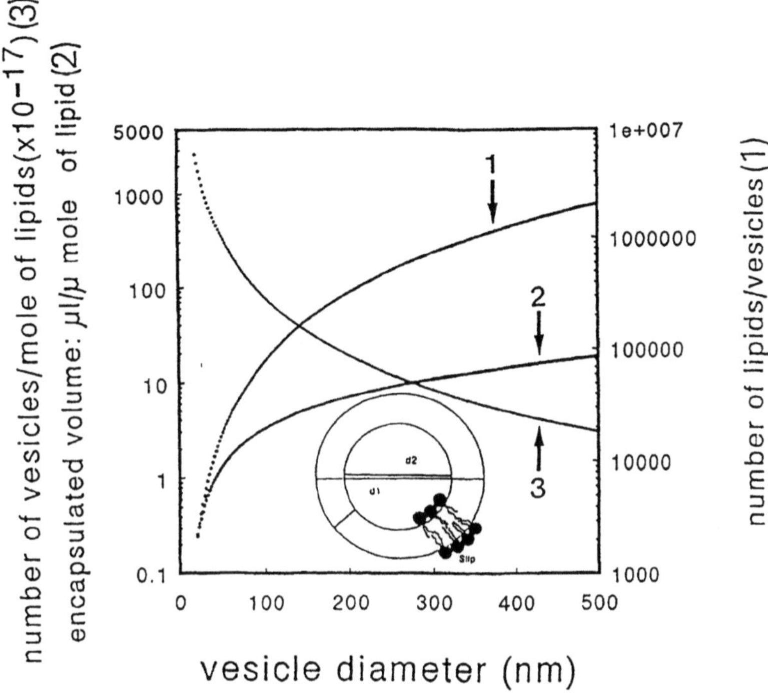

Fig. 3. Variations in the number of lipid molecules per vesicle, vesicle concentration and internal volume of the liposomes, as a function of the lipid concentration and the vesicle size (external diameter). The calculations have been done with d_1 and d_2, respectively, as external and internal diameters of the liposomes and assuming an area per phospholipid head group of $0.70\,nm^2$. $d_1 = d_2 + 2r$, with r (thickness of the bilayer) = 4 nm

high lipid concentration must be preferred for such an encapsulation. However, liposome applications may also impose the use of very small vesicles; for instance, drug delivery through endothelium may require the use of very small vesicles (Delattre 1993). Therefore, the choice of the technique required will depend on numerous criteria, which should be reviewed according to their specificities. A careful reading of Tables 1 to 3 and of the corresponding cited literature is highly recommended.

Among all these techniques, those using mechanical energy allow the highest lipid concentration use; in particular, the Bangham's techniques, extrusion through polycarbonate membrane, French press, microfluidiser and freeze thaw techniques are the most productive. One limitation in the microfluidiser, extrusion through polycarbonate membrane, and French press techniques is that they require sophisticated and rather expensive apparatus. However, compared to Bangham's and freeze-thaw techniques, they allow the preparation of comparatively well-calibrated liposomes.

For encapsulation of hydrophobic and/or amphipathic molecules, the above considerations are not valid. As these molecules concentrate into lipids, their encapsulation will depend only on the bilayer concentration. Then, they should be dissolved with the lipids by co-solubilisation in either an organic solvent or a detergent solution. Organic solvents very often denature molecules, and protein in particular (except in some specific cases, bacteriorhodopsin; Rigaud et al. 1983) and therefore, most of the time the processes using detergent are particularly preferred for protein incorporation.

Note. In some cases, giant liposomes with diameters of above $1\,\mu m$ are also required. These liposomes can be used to perform patch-clamp experiments, to measure surface tension or elasticity by optical microscopy. However, such kinds of liposomes are prepared using very special techniques, the description of which is outside the scope of this chapter.

4   Description of Four Selected Techniques of Liposome Preparation

The procedure of lipid mixing, which is common to all the preparations, is described below (Sect. 4.1). This is followed by a description of the preparations of multilamellar vesicles

(MLV; Sect. 4.2.1), small unilamellar vesicles (SUV; Sect. 4.2.2), large unilamellar vesicles (LUV; Sect. 4.2.3) and very large unilamellar vesicles (VLUV; Sect. 4.2.4). Each of these three procedures of liposome preparation is described independently; however, as indicated in Fig. 4, some parts of the procedure are common to all. Finally, in Section 4.3, quality controls are presented.

Note. It should be noted that the lipids used in these four preparations allow vesicles to form at room temperature. Changing the lipids may lead to change in temperatures, because vesicles should always be formed above the transition temperature of the mixture.

4.1 Lipid Mixing

This step is common for different types of liposome preparation. The lipid mixing procedure using organic solvents is discussed, since most of the time the phospholipids are delivered in organic solvent (frequently in the form of chloroform solutions). If the lipid sample is supplied as a dry compound, the use of solvent is recommended for lipid mixing.

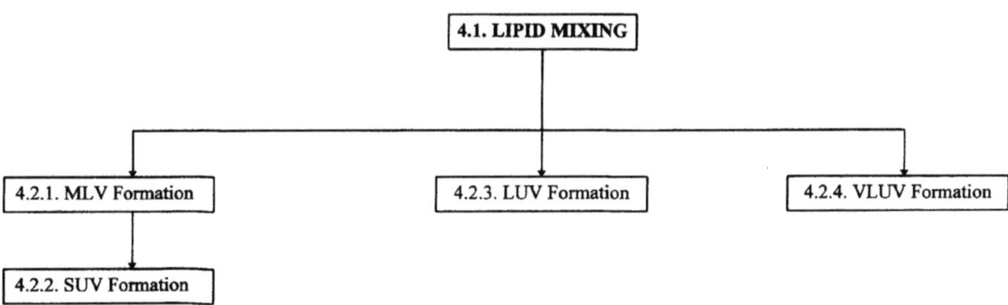

Fig. 4. Scheme of the different steps of liposome formation and of the common steps of each procedure. (The numbers in boxes refer to section numbers in text)

Equipment
- Vacuum pump
- Balance (with a precision of 0.01 mg)
- Laminar flow hood (when required for further experiments)
- Glass pipettes
- Scintillation vial (20 ml), a glass round bottomed flask (50 ml)

Chemicals   Depending on the types of vesicles required, different lipids are used:

- Egg phosphatidylcholine (EPC) in chloroform solution (100 mg/ml) (MW ~ 770) or equivalent lipid, or mixture of lipids forming lamellar phase at room temperature.
- Egg phosphatidic acid (EPA) in chloroform solution (10 mg/ml) (MW ~ 700).
- Cholesterol in chloroform solution (100 mg/ml)

Lipid mixing   The following mixtures of lipids are recommended for the different vesicles preparations:
For MLV, SUV and LUV: EPC:EPA, 9:1 molar ratio
For VLUV: EPC:EPS:cholesterol, 1:1:1 molar ratio

1. Clean the container (scintillation vial or glass round-bottomed flask) by keeping it in a sonication bath for 15 min in order to remove unwanted particles (this step is necessary where further use of sonication is envisaged in the protocol). Dry and weigh the container accurately.

2. Add chloroform solution of the first lipid to the container, for instance about 0.20 ml of EPC solution, i.e. about 20 mg of EPC.

3. Evaporate the chloroform under gentle nitrogen stream.

4. When no liquid chloroform is left, place the container under vacuum for 2 to 12 h.

5. Weigh the flask accurately. It can also be weighed after rehydration.

Note. Rehydration of PC is observed from the increase of !
weight following immediate transfer from vacuum to the balance. The weight increase would be about 1% of the PC.

6. Add the second lipid to the dried EPC film and repeat steps 3 to 5. If required, add the third component also and again repeat steps 3 to 5.

Note. Hydrophobic molecules to be incorporated in the !
membrane such as fluorescent or radiolabelled lipids, synthetics peptides, pharmaceuticals etc. are added at this stage using the same procedure as for EPA addition, provided they are not affected by contact with organic solvent in use. Only in very few cases may protein also be mixed to the lipids at this stage.

7. Calculate the molar ratio of each lipid from the weight of each lipid.

Note. Exact molecular weight of natural glycerolipids intro- !
duced may easily be deduced from the gas liquid chromatographic analysis of methyl esters of their fatty acids (Kates 1986).

8. After weighing, if the sample is not used immediately, fill the flask with argon (in order to avoid lipid degradation by oxidation) and close it.

Note. It is possible to store the dry preparations for long !
periods, for instance after grouping the preparations. In this case, scintillation vials (also used for MLV formation) may be replaced with a glass round-bottomed flask during the last solvent evaporation step, the sample being simply dried under nitrogen stream, and placed under vacuum.

Storage under argon at $-80\,^{\circ}\mathrm{C}$, in capped vials covered with Parafilm, easily provides ready to use samples. UV filtering glass flask may be used especially with very unsaturated samples.

4.2 Formation of Vesicles

4.2.1 Multi-Lamellar Vesicles (MLV)

Equipment
- Vortex mixer
- Scintillation vials (20 ml)

Chemicals Prepare the aqueous buffer (10 mM HEPES buffer, pH 7.4; 145 mM NaCl; 0.003% sodium azide, w/v) in advance, by filtering on 0.45 µm filter and store at 4 °C till further use.

Formation of MLV

1. Prepare the aqueous solution for hydrating the EPC/EPA dried film by dissolving all the components to be encapsulated in the internal compartment of liposomes. Pay attention to the osmolarity of the solution.

2. Add exactly 3 ml of this solution (or weigh it after addition) to the dried EPC/EPA film.

3. Vortex the mixture gently for about 5 min at room temperature until the lipid film disappears from the wall of the flask. The dispersion must appear white and homogeneous.

4. Eliminate the non-encapsulated hydrophilic molecules from the MLV preparation. Centrifuge the sample in a Beckman 50-Ti rotor at 100 000 g for about 1 h. Remove the supernatant and resuspend the pellet in the initial volume. This is the most easy and convenient method of removing non-encapsulated molecules.

5. Transfer the sample to an argon-filled 20-ml glass vial, wrap in aluminium foil to protect in from light damage, and store at 4 °C.

4.2.2 Small Unilamellar Vesicles (SUV)

For almost all the liposome procedures using mechanical energy, the formation of MLV is an important prerequisite. The final size of the unilamellar vesicles formed depends on the procedure chosen. Therefore, the steps described above of "lipid mixing" and "Formation of Multilamellar Vesicles" (except for the elimination of non-encapsulated molecules and the quality control) should be completed before preparing SUV.

Equipment

- Sonication apparatus with $1/2''$ titanium probe mounted on a 500-W transducer head (Vibracell Sonics Sonifier)
- Balance (with a precision of 0.01 mg)
- Table-top centrifuge
- Centrifuge tubes
- Pipetman (1 ml)
- Millex filters (Millipore)
- Scintillation vial (20 ml) (Polylabo, Strasbourg, France)
- Sephadex G-25 column (PD 10 from Pharmacia)

Formation of SUV

1. Place the EPC/EPA MLVs in an ice-cold water bath (4 °C) and sonicate by using a probe terminated with a $1/2''$ titanium tip placed as indicated in Fig. 5.

2. Centre the probe perfectly in the flask, to avoid touching the wall of the vial (Fig. 5).

3. Introduce nitrogen gas laterally into the vial through a catheter of 1 mm external diameter, in order to avoid oxidative degradation during sonication. Keep the nitrogen flow between 20–30 ml/min in order not to allow the aerosols formed to leave the vial and to avoid chemical loss. Keep the flow of the gas away from the titanium probe (see Fig. 5).

4. Sonicate the dispersion five times (each time for 2 min periods with 1-min interruptions between each phase; this regimen of sonication is valid for a 20 mM lipid concentration. This duration should be increased as a function of lipid concentration). During the sonication period (burst), the lipid is strongly agitated by ultrasonication but it should not be oscillating in the flask. At least 1 min interval is required

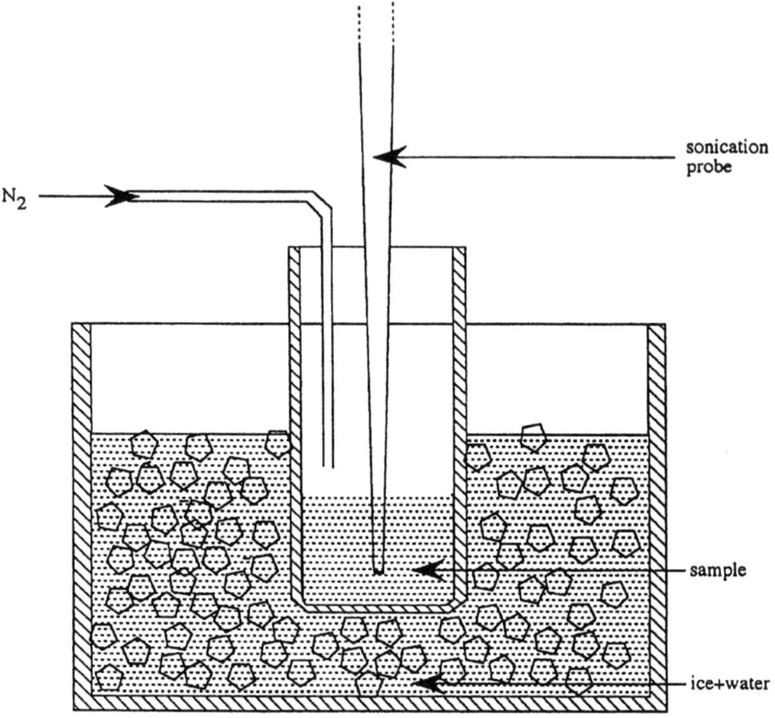

Fig. 5. Setup for the preparation of SUV by sonication. For high ultrasound power input, the sonication probe schemalized on this drawing is replaced by the 1/2" probe described in the text.

for the sample to cool down (sonication increases the temperature of the sample, that may increase the rate of hydrolytic degradation).

! **Note.** As sonication efficiency is highly variable and depends on parameters which can easily be adjusted, it is recommended to determine them before the use of sonication and apparatus in order to avoid sonication problems, for reproducibility.

In this respect we recommend, after tuning the ultrasound power supply (according to manufacturer's recommendations), adjusting the maximal input of ultrasounds in the sample. To this effect, a vial identical to that which will be used for sonication is filled with the same amount of water and placed under conditions similar to those used for sonication, except the use of an ice bath. Then, by successive adjustments, the depth to which the sonication tip is immersed is progressively adjusted by measuring the temperature reached by the water between each burst of precisely about 1 or 2 min. The tip position which gives the highest increase of temperature corresponds to the best yield.

5. Wipe the sonicator probe at the end of sonication, laterally with the neck of the vial, in order to remove drops of the solution left on it when pulled out of the dispersion. The dispersion appears transparent like water and light blue (Tyndall effect) if no chemical has modified the solution.

6. Quantitatively transfer the preparation into a centrifuge tube and centrifuge at 2000 g for 10 min, to remove titanium particles coming from the tip by abrasion.

7. After centrifugation, filter the sample through a 0.22-µm Millex filter in order to sterilise the sample.

Note. Depending upon the nature of lipids, some absorption ! and consequently loss of material may occur during this filtration step.

8. Remove the non-encapsulated molecule, by passing the sample through a PD10 column (see Quality Control, Sect. 4.3, Encapsulation).
   - Equilibrate the column with 200-ml buffer free of the encapsulated molecule.
   - Place 2.5 ml of the sample on top of the column.
   - Add 3.5 ml of buffer when all the sample has percolated into the column. Collect simultaneously, the eluate in a scintillation vial.

9. Store the sample as described in step 5 of MLV preparation (Sect. 4.2.1).

4.2.3 Large Unilamellar Vesicles (LUV)

As mentioned earlier, all the steps of "lipid mixing" should be completed before LUV preparation.

Equipment
- Vortex mixer
- Sonication apparatus (bath or sonication probe)
- Rotary evaporator
- Manometer
- Water bath for the rotatory evaporator
- Round-bottomed flask (50 ml)
- Polycarbonate membranes: pore sizes: 0.8, 0.4, 0.2, 0.1 and 0.05 μm
- Filter holder (e.g. Lipofast Basic apparatus from Avestin, Leiden, The Nertherlands)
- 20-ml scintillation vials (Polylabo, Strasbourg, France)

Chemicals
- Ethyl ether (freshly distilled)
- Prepare the aqueous buffer (10 mM HEPES buffer, pH 7.4; 145 mM NaCl; 0.003% sodium azide, w/v) in advance, by filtering on 0.45 μm filter and store at 4 °C till use.
- Any hydrophilic molecule to be encapsulated must be introduced in the buffer used for the preparation.

Formation of LUV

1. Dissolve the dry EPC/EPA film, possibly containing the hydrophobic molecules to be incorporated in the liposome membranes (e.g. fluorescent probes, and in some very particular cases the membrane protein) in the round bottom flask in 3 ml of ethyl ether.

! **Note.** Use gentle vortexing in order to remove all the lipids from the flask wall.

2. Add 1 ml of the aqueous buffer solution as soon as all the lipids are dissolved.

3. Sonicate the suspension in order to mix the two immiscible phases, with the sonication probe for 30 s or for 2–3 min in a bath sonicator. The emulsion formed must be stable for

at least 10 min. If this is not the case, size heterogeneity of the vesicles results in the final preparation.

4. Place the round-bottomed flask containing the suspension on the rotary evaporator immediately after the sonication. Rotate the flask gently (30–40/min) in a water bath set at about 20 °C, so that the evaporation, which must be controlled by nitrogen flux, is very slow (the pressure in the evaporator is about 460 mm Hg; 0.6 Atm).

5. Note that during the ethyl ether evaporation, an intermediate "gel-like" phase is formed. Upon further evaporation, this would result in the liquidisation of this gel phase. Wait for the complete liquidisation of the gel phase, which may take from 15 to 25 min.

6. Remove vacuum and slowly fill the evaporator with the nitrogen gas up to atmospheric pressure.

7. Add 2 ml of the aqueous solution prepared above and gently vortex it in order to completely homogenise the suspension (particularly for the small amount of gel phase which might have been left).

8. Continue the evaporation of ethyl ether, but more rapidly at higher vacuum (100 mm Hg; 0.13 Atm), in order to completely eliminate the organic solvent (this step can take 45 to 60 min).

9. Quantitatively transfer the liposome suspension to a clean and weighed 20-ml scintillation vial. Weigh the suspension in order to have the precise measurement of the final volume of the sample (assuming a density of 1.0).

10. Using a filter holder pass the liposomes through the successive polycarbonate membranes of 0.8, 0.4, 0.2, 0.1 and 0.05 µm to obtain an accurate calibration of the vesicles.

Note. An apparatus (Lipofast), which consists of two identical !
glass syringes of 1 ml (or larger volumes when necessary) sepa-

rated by a filter holder, can conveniently be used rather than a simple filter holder since it allows back and forth extrusions through the same membrane. The sequential use of the five different-sized membranes allows obtaining monodisperse and stable large unilamellar vesicles of 120 nm (Lesieur et al. 1991).

11. Remove the non-encapsulated molecules on a PD10 column as indicated in step 8 of SUV formation (Sect. 4.2.2).

12. Store the liposomes as described in step 5 of MLV formation (Sect. 4.2.1).

4.2.4 Very Large Unilamellar Vesicles (VLUV)

Equipment
- Vortex mixer
- Water bath
- Dialysis bags (1 cm diameter)

Chemicals
- PC, PS and cholesterol
- Biobeads SM-2 from Bio-Rad (Richmond, California, USA)
- n-Octyl β-D-glucopyranoside (Sigma, St. Louis, MO, USA)
- Prepare the aqueous buffer (10 mM HEPES buffer, pH 7.4; 145 mM NaCl; 0.003% sodium azide, w/v) in advance, by filtering on 0.45 μm filter and store at 4 °C till use.
  Any hydrophilic molecule to be encapsulated must be introduced in the buffer used for the preparation.

! **Note.** The following procedure is described for a 1:1:1 molar ratio of PC:PS:cholesterol mixture. It gives vesicles with a mean diameter of 0.85 μm. Preparation with other lipid mixture would give vesicles of different mean diameter.

Formation of VLUV
1. Hydrate the dry EPC/PS/ cholesterol film (20 μmol) by 0.125 ml of buffer, for 30 min at room temperature after vortex mixing.

2. Add 58.5 mg of n-Octyl β-D-glucopyranoside (200 µmol).

3. Adjust the volume of the sample to 0.625 ml with the buffer.

4. Shake the mixture vigorously.

5. Transfer the sample to a dialysis bag and dialyse it for 21 h under gentle stirring against 100 ml of buffer containing 2 g of Biobeads.

6. Remove the dialysis buffer, and dialyse the sample again for 1 h against the desired buffer.

4.3 Quality Controls

Determine the phospholipid concentration either by phosphorus assay (as described in Chap. IV, this Vol.) or by using enzymatic assay. The enzymatic assay is more specific and less sensitive to impurities. **Lipid assay**

Note. For phospholipids assay using phospholipase D, choline oxidase and peroxidase and for cholesterol assay using cholesterol esterase, cholesterol oxidase and peroxidase, kits are available (Biomérieux, Mary-L'Etoile, France). **!**

The size stability of the preparation can be followed by quasi-elastic light scattering measurements (it directly gives the average diameter of the vesicle when unaggregated) and by the polydispersity parameters of the sample. It can also be followed by gel exclusion chromatography using a double detection (UV-Vis + refractive index detection) when vesicles aggregation is suspected (Lesieur et al. 1993). Turbidity measurements performed on a spectrophotometer are also adequate but the influence of size variations on optical density measurement may be difficult to interpret in certain size ranges (Lesieur et al. 1993). **Physical stability**

Encapsulation In order to test the retention of the encapsulated molecules versus time, the step 4 of MLV preparation (Sect. 4.2.1) can be repeated at different times following the preparation. The molecules remaining in the supernatant can be assayed by appropriate techniques. For unilamellar vesicles (from SUV to LUV), different types of gel exclusion chromatography from very simple exclusion on Sephadex G25 (Pharmacia); (see Sect. 4.2.2, step 8) to more complex gel exclusion chromatography using either UV-Vis, refractive index, radiolabelled molecule or fluorescence detection can also be applied to test the retention of encapsulated molecule. A convenient and low-cost version of this technique is described below.

Gel exclusion high performance liquid chromatography Different types of gel exclusion chromatography have been developed for liposome quality control. These techniques, which allow sizing and polydispersity measurement as well as encapsulation efficiency and release controls, are compared in Lesieur et al. (1993). The basic equipment depicted in Fig. 6 allows vesicle sizing. In addition, gel exclusion chromatography of vesicles easily allows their separation from non-encapsulated materials and therefore evaluation of encapsulation yield. While a basic setup is described below, detailed recommendations for the use of the technique are given in Ollivon et al. (1986); Lesieur et al. (1993); Walter et al. (1993).

The basic gel exclusion chromatography system depicted in Fig. 6 is assembled from low pressure liquid chromatography equipment:

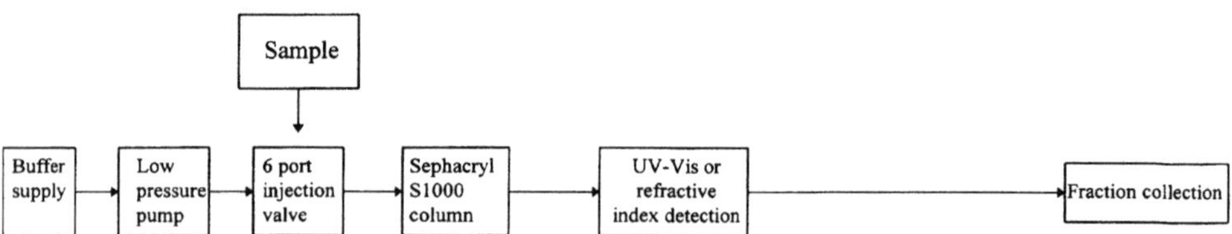

Fig. 6. Schematic diagram of the gel exclusion chromatography

- A pump able to deliver a constant flow (about 0.3 ml/min) of aqueous buffer.
- A 6-port injection valve (rotating Teflon valve, Type 50 Rheodyne, Cotati, CA, USA) equipped with a 0.1- to 1-ml injection loop, the volume of which is chosen as a function of sample concentration and specific turbidity.
- A glass column (1 × 20 cm, Pharmacia) filled with Sephacryl S-1000 (Pharmacia) and saturated with lipids.
- An UV-Vis or refractive index detector connected to a recorder.
- A fraction collector, the collected fractions of which might be subsequently analysed by external means (UV-Vis spectrometer, radioactivity or fluorescence measurements, etc.). While refractive index or UV detection allows detection of SUVs at a concentration corresponding to 1-μM phospholipids or less at the lowest wavelengths, this limit may easily be overcome for larger vesicles.

5 Guidelines and Strategies for Membrane Protein Reconstitution

Among all the techniques available to prepare liposomes, those using detergents are the most efficient to reconstitute "proteoliposomes". In order to reconstitute a membrane protein into artificial membranes, the first step is its purification, which can be achieved by solubilising the natural membrane by detergents. Moreover, the broad variety and choice of detergents (amphiphilic molecules) available make them appropriate to maintain the structure and activity of the protein, even when it is extracted from its natural membrane. However, most of the methods developed for proteoliposomes reconstitution using detergents are empirical and require numerous and tedious experiments to find some reproducible procedures of the reconstitution. In 1988, Paternostre et al. and Rigaud et al. have proposed a new strategy for the incorporation of bacteriorhodopsin, which can be generalised to any membrane protein and has already been successfully applied for the reconstitution of other proteins (Levy et al. 1990).

5.1 Liposome Reconstitution Using Detergent

The mechanism by which very large aggregates such as liposomes are formed by detergent elimination from a homogeneous suspension of mixed micelles is called micelle to vesicle transition. This transition is reversible. In fact, the transition is not unique, as the system has to go through different stages and physical phases to reorganise into either liposomes or micelles. Different studies of this transition, performed on a variety of systems, have shown the complexity of the molecular and the supramolecular processes involved, which depend on numerous parameters of the systems studied such as lipid and detergent nature, lipid and detergent concentrations, rate of elimination/addition, temperature etc. (Ollivon et al. 1988; Lesieur et al. 1990; Seras et al. 1992, 1993).

However, from the comparison of the mechanisms, it appeared that some steps are common to all the transitions, the knowledge of which has made it possible to develop strategies of reconstitution. The micelle to vesicle transition is in four successive stages (Fig. 7). Addition of detergent into a suspension of liposomes results, as the detergent concentration is increased, in:

- Stage 1: at low concentration of detergent (as compared to lipid), the detergent partitions between the lipidic phase and the water without formation of mixed micelles. The incorporation of detergent into the membrane includes swelling of the liposomes and change in their morphology.
- Stage 2: further detergent incorporation includes a drastic increase in membrane permeability which can be related to

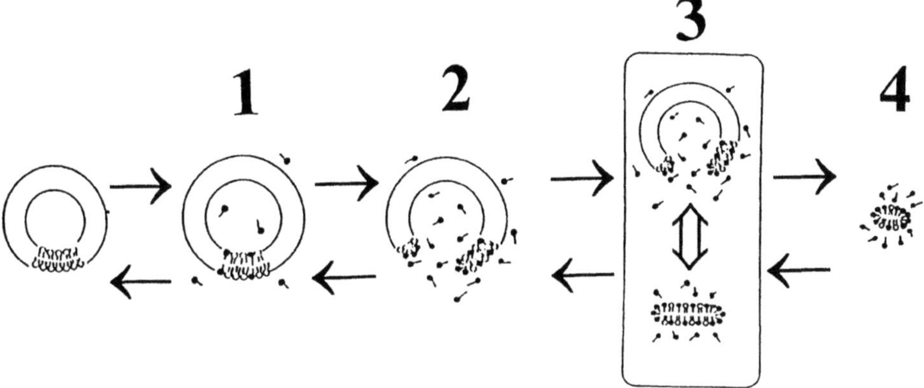

Fig. 7. Different steps of the micelle to vesicle transition

local detergent concentration and liposomes openings. At this stage, no mixed micelles are formed.

- Stage 3: this stage is characterised by the coexistence of detergent saturated membrane with lipid saturated mixed micelles. Increasing the detergent concentration will progressively cause the disappearance of the lamellar phase in favour of the mixed micellar one.

- Stage 4: this stage is characterised by the mixed micellar structure. Increasing the detergent concentration will result in decrease in size of the mixed micelles and the dilution of the lipids into these micelles.

The transitions between each of these stages are strictly dependent on the detergent to lipid molar ratio in the aggregates and on the monomeric detergent concentration in equilibrium with these aggregates. For each lipid and detergent system, it is important to determine the boundaries between these phases. The simplest way to determine these limits is to record the turbidity variations of the system during the transition. The turbidity measurement is a function of the size, shape, concentration and refractive index of the aggregates in solution. As a

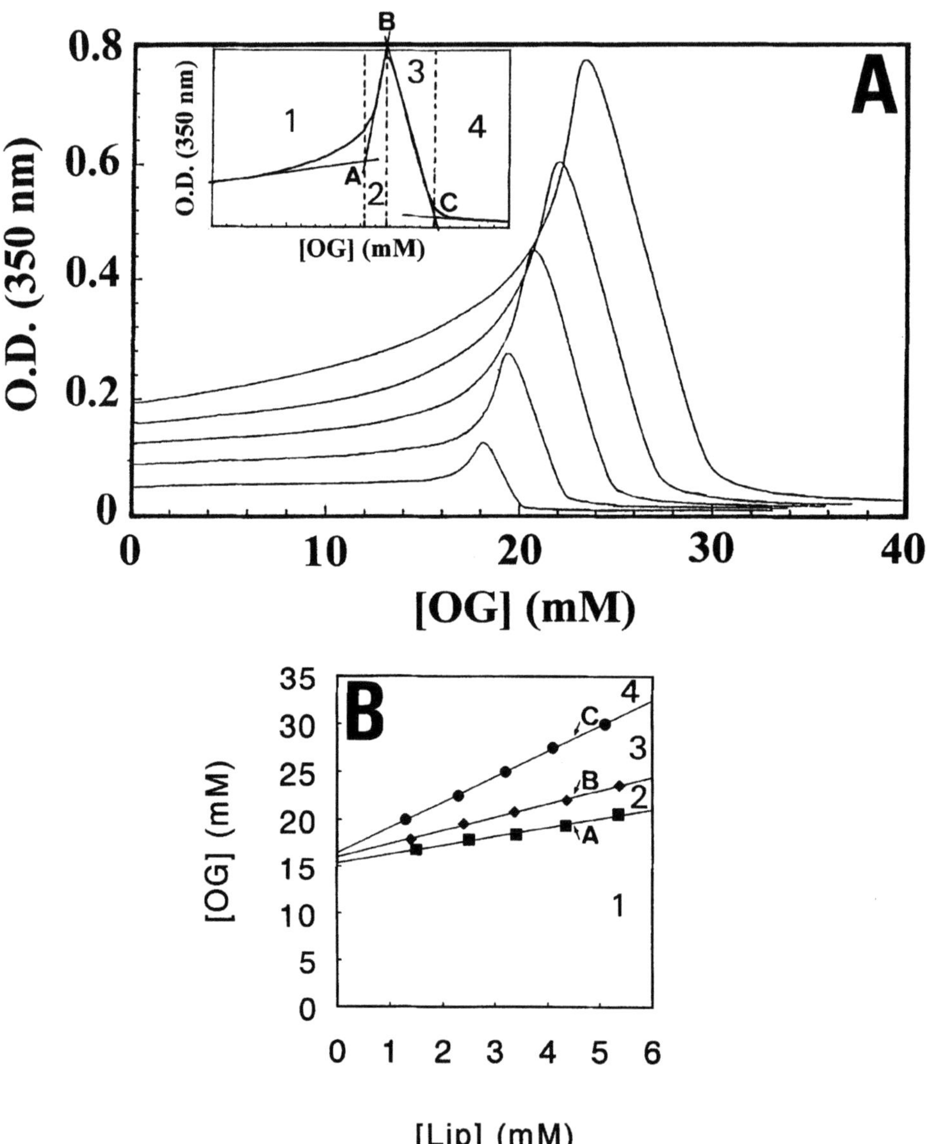

Fig. 8. Determination of the boundaries of the vesicle-micelle transition of egg yolk phosphatidylcholine (EPC) and octyl glucoside (OG). A turbidity profiles (OD at 350 nm) obtained during the solubilization of EPC vesicles by OG. The five curves were obtained for five different initial lipid concentrations (*from left to right*) 5.5, 4.5, 3.5, 2.5 and 1.5 mM. *Insert* Graphic determination of the break points *A*, *B* and *C* delimiting stage 1, 2, 3 and 4 of the solubilisation process (see Fig. 7). **B** The total OG concentrations required to reach the breakpoints *A*, *B* and *C* has been measured on each curves of Fig. 8A and reported as a function of the total lipid concentration in the cuvette. The equations of the linear relation ships obtained between OG and lipid concentrations are:

breakpoint A: $[OG]_A = [OG]0_A + (R_A \cdot [Lip]_A)$
breakpoint B: $[OG]_B = [OG]0_B + (R_B \cdot [Lip]_B)$
breakpoint C: $[OG]_C = [OG]0_C + (R_C \cdot [Lip]_C)$

$[OG]_{A,B,C}$: total OG concentrations required to reach the breakpoints
$[OG]0_{A,B,C}$: OG concentrations extrapolated from the straight lines at $[Lip] = 0$. These concentrations represent the monomere OG concentrations, i.e. the concentrations of OG which are not in interaction with lipids, in equilibrium with the mixed structures formed at the boundaries
$R_{A,B,C}$: slopes of the straight line, which represent the molar ratios between OG and lipids in the structures present at the boundaries. These ratios are expressed by the following relations:
$R_A = [OG]i_A/[Lip]_A$ $R_B = [OG]i_B/[Lip]_B$ $R_C = [OG]i_C/[Lip]_C$
$[OG]i_{A,B,C}$: OG concentrations in interaction with the lipids
$[Lip]_{A,B,C}$: total lipid concentrations (the monomere concentration of the lipid is negligible)

Finally, these straight lines delimit the existence of the 4 different stages of the transition (stage 1, 2, 3 and 4, see Fig. 7).

For the EPC-OG systems, the phases boundaries are given by the following relations:

breakpoint A: $[OG]_A = 15.3 + 0.92 \cdot [Lip]$
breakpoint B: $[OG]_B = 16.1 + 1.36 \cdot [Lip]$
breakpoint C: $[OG]_C = 16.6 + 2.57 \cdot [Lip]$

result, these measurements are extremely sensitive to the transition. The turbidity profiles obtained for different lipid concentrations do not change in shape, but are shifted along with the detergent concentration axis, indicating the dependence of the phenomenon on the concentration of each compound.

These profiles also exhibit reproducible breakpoints, which have been interpreted as the limits of the liposome opening/. closure (point A), as the onset (point B) and the offset (point C) of the micelle/vesicle transformation (Fig. 8). The detergent dependence of these limits on lipid concentration is linear. From the slope of the straight line the detergent to lipid concentration ratio in the aggregate, and from the intercept with the detergent axis at zero lipid concentration, the monomeric detergent concentration in equilibrium with these aggregates are determined. The determination of these limits is essential for the following reconstitution strategy.

5.2   Strategy for the Determination of Optimal Conditions of Membrane Protein Incorporation into Liposomes Using Detergents

Artificial and natural membranes are organised molecular systems. Knowledge of how the different molecules interact together is required to reconstitute well-defined systems. The strategy proposed is summarised in Fig. 9.

The essential strategy of the method proposed involves the addition of solubilised protein to a suspension containing lipid and detergents in stages 1, 2, 3 or 4 of the transition and then to analyse the proteoliposomes obtained after detergent removal. The comparison between the different reconstituted systems allows determining the optimal initial conditions of lipid-protein and detergent mixing. Detailed experimental procedure is given for the reconstitution of the membrane protein bacteriorhodopsin (BR).

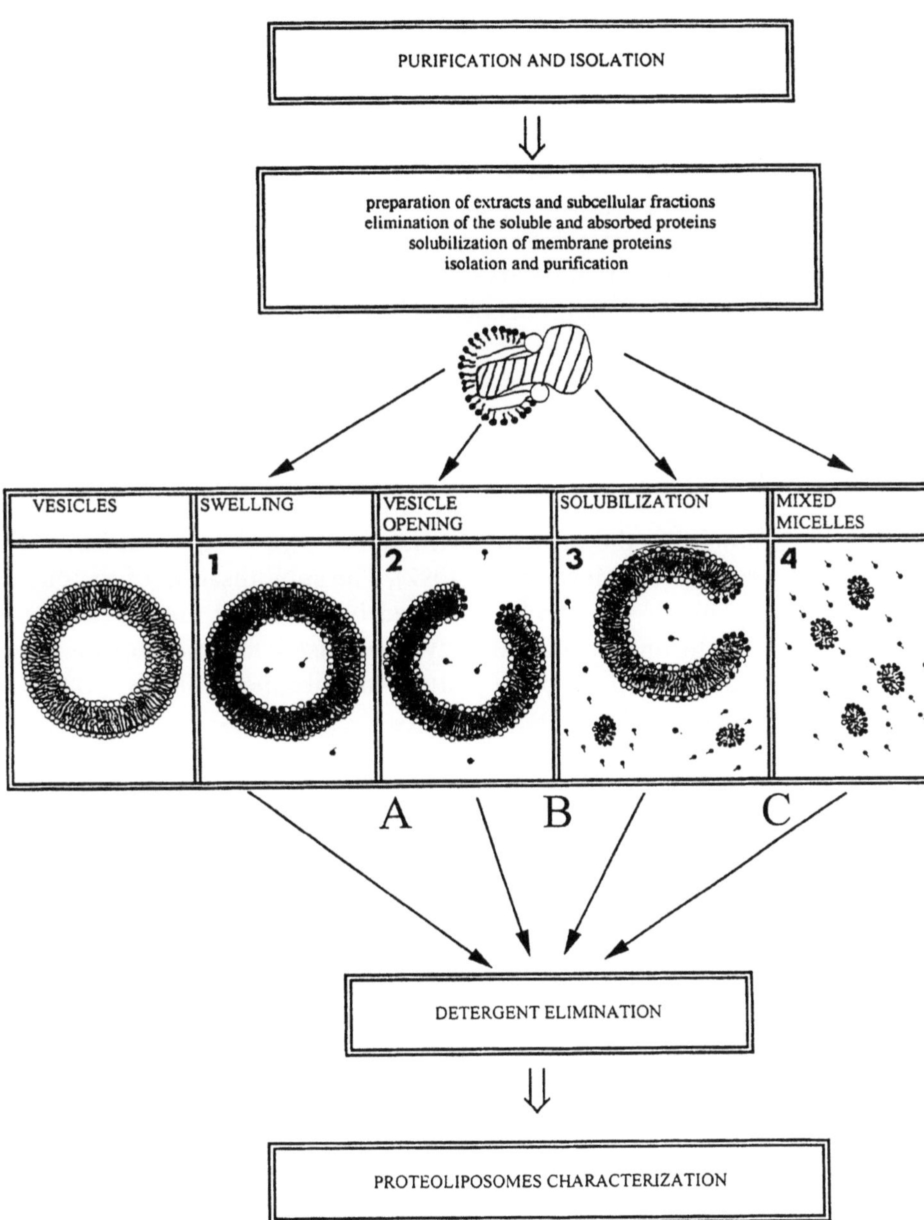

Fig. 9. Strategy for protein incorporation into liposomes

Equipment
- UV-visible spectrophotometer
- Glass cuvette
- Pipetman (0.1 ml)

Materials
a) Buffered solution (20 mM KH_2PO_4, 110 mM K_2SO_4, pH 7)
b) EPC/EPA LUV prepared as described above (see Sect. 4.2.3) but in buffer defined above (a).
c) Concentrated detergent solutions prepared in buffer (a):
   50 mM sodium cholate(Na-Chol): (MW = 430, CMC = 2–10 mM)
   200 mM octyl glucoside(OG): (MW = 292; CMC = 21 mM)
   30 mM Triton X-100 (TX-100): (MW = 650; CMC = 0.1 mM)

Solubilisation of the natural membrane (purple membrane) containing the protein (BR) is done according to established procedures (Meyer et al., 1992). The solubilisation of the purple membrane can be followed by turbidity measurements. This technique has already been used for monitoring the solubilisation of other natural membrane proteins (Kragh-Hansen et al. 1993).

Determination of the boundaries between stages 1 to 4

1. Prepare six vesicle suspensions (2 ml each) at different lipid concentrations (0.5, 1, 1.5, 2, 2.5 and 3 mM).

2. Weigh the volume added in a glass cuvette for each lipid concentration, in order to have the initial volume.

3. Measure the optical densities at 400 nm (or any wavelength for which no absorption occurs) of the initial vesicles suspension and after each successive addition of aliquots (5 µl) of detergent solution (Na-Chol, OG or TX-100).

4. Repeat the addition until OD decreases (0.01) and remains constant, after further addition of detergent.

5. Repeat steps (2) to (4) for each sample with different lipid concentrations.

6. Plot OD versus the detergent concentrations on the same graph for each sample.

Note. The lipid and detergent concentrations should be corrected from the dilution factor brought about by the detergent addition.

7. Determine, according to Fig. 8, lipid and detergent concentrations at each of the breakpoints A, B, and C and plot detergent concentration versus lipid concentration. Determine the detergent/lipid molar ratio in aggregate at each breakpoint (slopes of the linear curve) and the monomeric detergent concentration in equilibrium with this aggregate (origin of linear curve). From this linear relationship, bring the detergent and lipid system in phase 1, 2, 3 or 4 at any lipid concentration.

1. All steps are identical to those described in previous paragraphs, except for the beginning of the experiment (before the successive addition of detergent), add solubilised protein and measure OD. Add successive aliquots of detergent to the protein-containing suspension.

2. Stop addition of detergent in stage 1, 2, 3 or 4 of the solubilisation process.

3. Eliminate the detergent from these samples.

Determination in the presence of solubilised protein

5.3 Elimination of the Detergent

The elimination technique will depend on the nature of detergent. For Na-Chol the elimination is achieved by dialysis, for TX-100 and OG the elimination is achieved by direct contact with Biobeads SM2 (Biorad, Hercules, CA, USA).

5.3.1 Elimination of Sodium Cholate by Dialysis

- Dialysis membranes (4–6 mm diameter, molecular weight cut-off: 12 000–14 000 Da, Spectrapor, Houston, TX, USA)
- Magnetic stirrer
- 1-l glass flask

Equipment

Elimination of sodium cholate

1. Prepare four dialysis membranes of 6 cm length.

2. Soak them for at least 10 min in the buffer used for the experiment.

3. Tightly close one end of the membrane, using a special clamp.

4. Put sample in wet dialysis membrane.

5. Tightly close the other end of membrane.

6. Put dialysis bag in 1 l of buffer used for the experiment.

7. Leave sample under stirring for at least 12 h at 4 °C.

5.3.2 Elimination of Triton X-100 and Octyl Glucoside

Equipment

- Biobeads SM2 (Biorad, Hercules, CA, USA)
- Magnetic stirrer
- 20-ml glass scintillation vials

Biobeads washing (Holloway 1973)

1. Weigh about 10 g of dried Biobeads SM2.

2. Add 500 ml of methanol and stir for 30 min.

3. Remove methanol.

4. Repeat steps 1 to 3 at least three more times.

5. Add 1 l of water (deionised, filtered through 0.22-μm filter and degassed) and stir for 30 min.

6. Remove water.

7. Repeat steps 5 to 6 at least three more times.

8. The washed Biobead SM2 can be stored at room temperature, by protecting from light for 1 month in deionised and degassed water.

! **Note.** The Biobeads can be recycled using same procedure.

1. Prepare four samples containing 5 mM of lipids, the membrane proteins and one of the four TX-100 concentration required to bring the systems to stage 1, 2, 3 or 4 of the solubilisation process.

Elimination of Triton X-100 (Levy et al. 1990)

2. Add 80 mg/ml of wet Biobeads SM2.

Note. In order to keep Biobeads SM2 active, it is necessary to keep them away from air. To weigh the Biobeads SM2, the water is removed with a Pasteur pipette: they are very quickly weighed and resuspended in solution containing detergent to be removed. **!**

3. Keep sample under gentle magnetic stirring for 3 h.

4. Add 80 mg/ml of fresh Biobeads SM2.

5. Keep sample under gentle magnetic stirring for 2 additional hours.

6. Isolate sample from the Biobeads SM2 using a Pasteur pipette and store at 4 °C.

Note. The capacity of Biobeads SM2 to absorb TX-100 is 180 mg of TX-100/g of Biobeads SM2. The capacity of Biobeads SM2 to absorb phospholipids is 0.9 to 2.6 mg phospholipids/g of Biobeads SM2 if they are organised in liposomes or dissolved in micelles, respectively. **!**

1. Prepare four samples containing 5 mM of lipids, the membrane proteins and one of the four OG concentrations required to bring the systems in stage 1, 2, 3 or 4 of the solubilisation process.

Elimination of Octyl glucoside

2. Add 80 mg/ml of wet Biobeads SM2.

Note. In order to keep Biobeads SM2 active, it is essential to keep them away from air. To weigh the Biobeads SM2, the water is removed with a Pasteur pipette; they are rapidly weighed and immediately resuspended in the solution containing the detergent to be removed. **!**

3. Keep sample under gentle magnetic stirring for 3 h.

4. Add 80 mg/ml of fresh Biobeads SM2.

5. Keep sample under gentle magnetic stirring for 2 additional hours.

6. Isolate sample from the Biobeads SM2 using a Pasteur pipette and keep at 4 °C.

! **Note.** The capacity of Biobeads SM2 to absorb OG is 300 µmol/ g of Biobeads SM2. The capacity of Biobeads SM2 to absorb phospholipid is 0.9 to 2.6 mg of phospholipids/g of Biobeads SM2 if they are organised in liposomes or dissolved in micelles, respectively.

5.4   Quality Controls of the Proteoliposomes

After detergent elimination, the resulting proteoliposomes are analysed and characterised for different criteria which essentially are:

- Size and polydispersity: by electron microscopy and gel exclusion chromatography
- Structure (uni- or multi-lamellar): by electron microscopy
- Surface Density of protein in proteoliposomes: by electron microscopy and sucrose gradient
- Homogeneity of protein preparation: by electron microscopy and sucrose gradient
- Activity of protein: by specific activity determination
- Efficiency of reconstituted protein as compared to that of protein in situ
- Orientation of the protein: by proteolytic cleavage and non-permeant specific inhibitors
- Membrane permeability: by fluorescent probes

5.5 Conclusion

The strategy described above allows determination of the optimal conditions for the reconstitution. It has been shown in previous studies that, for a given membrane protein (bacteriorhodopsin) and given lipids (EPC:EPA, 90:10%), the optimal reconstitution depends on the chosen detergent. In particular, three detergents were tested: Octyl β-D-glucopyranoside (OG), Triton X-100 (TX-100), the sodium salt of cholic acid (Na-Chol). These detergents, frequently used for protein reconstitution, are either non-ionic (OG and TX-100) or ionic (Na-Chol), and contain pure molecules (OG and Na-Chol) or a mixture (TX-100).

The mechanism by which the protein is reconstituted exclusively depends on the detergent. With OG, a direct incorporation of BR is performed which resulted in detergent-saturated liposomes and in a well-oriented (95%) protein incorporation. The resulting liposomes were homogeneous in size and slightly smaller than the initial ones. With TX-100, BR interacted with detergent-saturated liposomes in the stage 2, by a protein exchange between mixed micelles and liposomes. The resulting reconstituted system was 80% oriented. Finally, with Na-Chol, BR was reconstituted from the totally solubilised systems and the protein was 70% oriented. The different behaviour of the detergent has been observed with other proteins as well, which demonstrates the importance of the lipidic organisation in the reconstitution procedures.

References

Bangham AD, Standish MM, Watkins JC (1965) Diffusion of univalent ions across the lamellae of swollen phospholipids. J Mol Biol 13:238–252
Barenholz Y, Amselem S, Lichtenberg D (1979) A new method for preparation of phospholipid vesicles (liposomes) – French press. FEBS Lett 99:210–213

Batzri S, Korn ED (1973) Single bilayer liposomes prepared without sonication. Biochim Biophys Acta 298:1015–1019

Blok MC, van Deenen LLM, de Gier J (1976) Effect of the gel to liquid crystalline phase transition on the osmotic behaviour of phosphatidylcholine liposomes. Biochim Biophys Acta 433:1–12

Bruner J, Srabal P, Hauser H (1976) Simple bilayer vesicles prepared without sonication: physico-chemical properties. Biochim Biophys Acta 455:322–331

Cullis PR, and Hope MJ (1985) Physical properties and functional roles of lipids in membranes. In: Vance DE, Vance JE (eds) "Biochemistry of lipids and membranes". Benjamin/Cumming Menlo Park, California, pp 25–72

Deamer D, Bangham AD (1976) Large volume liposomes by an ether vaporization method. Biochim Biophys Acta 443:629–634

Delattre J (1993) Liposomes et barrières endothéliales. In: Delattre J et al. (eds) Les liposomes: aspects technologiques et pharmacologiques. INSERM Paris, pp 167–176

Enoch HG, Strittmater P (1979) Formation and properties of 1000-Å-diameter, single-bilayer phospholipid vesicles. Proc Natl Acad Sci USA 76:145–149

Fettiplace R, Haydon DA (1980) Water permeability of lipid membranes. Physiol Rev 60:510–550

Grift M, Crommelin DJA (1993) Chemical stability of liposomes: Implications for their physical stability Chem Phys Lipids 64:3–18

Hamilton RL, Guo L (1984) French pressure cell liposomes: preparation, properties and potential. In: Liposome Technology CRC Press, Boca-Raton, pp 37–50 (Volume I)

Holloway PW (1973) Simple procedure for removal of Triton X-100 from protein samples. Anal Biochem 53:304–308

Hope MJ, Bally MB, Mayer LD, Webb G, Cullis PR (1985) Production of large unilamellar vesicles by a rapid extrusion procedure. Characterisation of size distribution, trapped volume and ability to maintain a membrane potential. Biochim Biophys Acta 815:55–65

Hope MJ, Bally MB, Mayer LD, Janoff AS, Cullis PR (1986) Generation of multilamellar and unilamellar phospholipid vesicles. Chem Phys Lipids 40:89–107

Horigome T, Sugano H (1983) A rapid method for removing of detergents from protein solution. Anal Biochem 130:393–396

Huang C (1969) Studies of phosphatidylcholine vesicles: formation and physical characteristics. Biochemistry 8:344–349

Ipsen JH, Mouritsen OG, Bloom M (1990) Relationships between lipid membrane area, hydrophobic thickness and acyl-chain orientational order: the effects of cholesterol. Biophys J 57:405–412

Kagawa T, Racker E (1971) Partial resolution of the enzymes catalyzing oxidative phosphorylation. J Biol Chem 246:5477–5487

Kashara M, Hinkle PC (1977) Reconstitution and purification of the D-Glucose transporter from human erythrocytes. J Biol Chem 252:7384–7390

Kates M (1986) Techniques of Lipidology: Isolation, analysis and identification of lipids. In: Burdon RH, van Knippenberg PH (eds) Laboratory Techniques in Biochemistry and Molecular Biology, Vol. 3 Part II. (2nd edn), Elsevier, Amsterdam

Kirby C and Gregoriadis G (1984) Dehydratation-rehydratation vesicles: a simple method for high yield drug entrapment in liposomes. Biotechnology 12:979–984

Kragh-Hansen U, Le Maire M, Noel JP, Gulik-Krzywicki T, Moller J (1993) Transitional steps in the solubilization of protein-containing membranes and liposomes by non-ionic detergent. Biochemistry 32:1648–1658

Kremer JMH, Esker MW, Pathmamanoharan C, Wieresema PH (1977) Vesicles of variable diameter prepared by a modified injection method. Biochemistry 16:3932–3935

Lecuyer H, Dervichian DG (1969) Structure of aqueous mixtures of lecithin and cholesterol. J Mol Biol 45:39–57

Lelkes P.I (1984) The use of French pressed vesicles for efficient incorporation of bioactive macromolecules and as drug carriers in vitro and in vivo. In: Gregoriadis G (ed) Liposome Technology, CRC Press, Boca-Raton, vol. 1 pp 51–65

Lesieur S, Grabielle-Madelmont C, Paternostre M, Moreau JM, Handjani-Vila RM, Ollivon M (1990) Action of Octyl glucoside on non-ionic monoalkyl amphiphile-cholesterol vesicles: study of the solubilization mechanism. Chem Phys Lipids 56:109–121

Lesieur S, Grabielle-Madelmont C, Paternostre M, Ollivon M (1991) Size analysis and stability of lipid vesicles by high performance gel exclusion chromatography, turbidity and dynamic light scattering. Anal Biochem 192:334–343

Lesieur S, Grabielle-Madelmont C, Paternostre M, Ollivon M (1993) Study of size distribution and stability by high performance gel exclusion chromatography. Chem Phys Lipids 94:57–82

Levy D, Bluzat A, Seigneuret M, Rigaud JL (1990) A systematic study of liposome and proteoliposome reconstitution involving Bio-Bead-mediated Triton X-100 removal. Biochim Biophys Acta 1025:179–190

MacDonald RI, MacDonald RC (1983) Lipid mixing during freeze-thawing of liposomal membranes as monitored by fluorescence energy transfer. Biochim Biophys Acta 735:243–251

Mayhew E, Lazo R, Vail WJ, King J, Green AM (1984) Characterisation of liposomes prepared using microemulsifier. Biochim Biophys Acta 775:169–174

Milsman MH, Schwendener RA, Weder HG (1978) The preparation of large single bilayer liposomes by a fast and controlled dialysis. Biochim Biophys Acta 512:147–155

Mimms LT, Zampighi G, Nosaki Y, Tanford C, Reynolds JA (1981) Phospholipid vesicles formation and transmembrane protein incorporation using Octyl glucoside. Biochemistry 20:833–840

Moriyama R, Nakashima H, Makimo S, Koga S (1984) A study on the separation of reconstituted proteoliposomes and unincorporated membrane proteins by use of hydrophobic affinity gels with special reference to band 3 from bovine erythrocytes membranes. Anal Biochem 139:292–297

Ollivon M, Walter A, Blumenthal R (1986) Sizing and separation of liposomes, biological vesicles and viruses by HPLC. Anal Biochem 152:262–274

Ollivon M, Eidelman O, Blumenthal R, Walter A (1988) Micelle-vesicle transition of Egg Phosphatidylcholine and Octyl glucoside. Biochemistry 27:1695–1703

Olson F, Hunt T, Szoka FC, Vail WJ, Papahadjopoulos D (1979) Preparation of liposomes of defined size distribution by extrusion through polycarbonate membranes. Biochim Biophys Acta 557:9–23

Oshawa T, Miura H, Harada K (1984) A novel method for preparing liposomes with a high capacity to encapsulate proteineous drugs. Chem Pharm Bull 32:2442–2445

Papahadjopoulos D, Kimelberg HK (1974) Phospholipid vesicles (liposomes) as models for biological membranes: their properties and interactions with cholesterol and proteins. In: Davison SG (ed) Progress in Surface Science. Pergamon Press, Oxford

Papahadjopoulos D, Miller N (1967) Phospholipid model membranes. I Structural characteristics of hydrated liquid crystals. Biochim Biophys Acta 135:625–638

Papahadjopoulos D, Vail WJ, Newton C et al. (1977) Studies on membrane fusion. III The role of calcium-induced phase changes. Biochim Biophys Acta 465:579–598

Paternostre M, Roux M, Rigaud JL (1988) Mechanisms of membrane protein insertion into liposomes during reconstitution procedures involving the use of detergent 1. Solubilization of large unilamellar liposomes (prepared by reverse phase evaporation) by Triton X-100, Octyl glucoside and sodium cholate. Biochemistry 27:2668–2677

Phillipot JR, Mustaftschiev S, Liautard JP (1983) A very mild method allowing the encapsulation of macromolecules into very large (1000 nm) unilamellar liposomes. Biochim Biophys Acta 734:137–143

Phillipot JR, Mustaftschiev S, Liautard JP (1985) Extemporaneous preparation of large unilamellar liposomes. Biochim Biophys Acta 821:79–84

Phillips MC (1972) The physical state of phospholipids and cholesterol in monolayers, bilayers and membranes. In: Danielli JF, Rosenberg MD, Cadenhead DA (eds) Progress in surface and membrane science, Vol 5. Academic Press, New York

Pick U (1981) Liposomes with a large trapping capacity prepared by freezing and thawing of sonicated phospholipid mixtures. Arch Biochem Biophys 212:186–194

Rhoden V, Golden SM (1979) Formation of unilamellar lipid vesicles of controllable dimensions by detergent dialysis. Biochemistry 18:4173–4176

Rigaud JL, Bluzat A, Buschlen S (1983) Incorporation of bacteriorhodopsin into large unilamellar liposomes by reverse phase evaporation. Biochem Biophys Res Commun 111:373–382

Rigaud JL, Paternostre M, Bluzat A (1988) Mechanisms of membrane protein insertion into liposomes during reconstitution procedures involving the use of detergent 2. Incorporation of light driven proton pump Bacteriorhodopsin. Biochemistry 27:2677–2688

Saunders L, Perrin J, Gammack DB (1962) Aqueous dispersion of phospholipids by ultrasonic radiations. J Pharm Pharmacol 14:567–572

Schwendener RA, Asanger M, Weder HG (1981) n-Alkyl-glucoside as detergents for the preparation of highly homogeneous bilayer liposomes of variable sizes (60–240 nm) applying defined rates of detergent removal by dialysis. Biochem Biophys Res Commun 100:1055–1062

Seras M, Handjani-Vila RM, Ollivon M, Lesieur S (1992) Kinetic aspects of the solubilization of non-ionic monoalkyl amphiphile-cholesterol vesicles by Octyl-glucoside. Chem Phys Lipids 63:1–14

Seras M, Ollivon M, Edwards K, Lesieur S (1993) Reconstitution of non-ionic monoalkyl amphiphile-cholesterol vesicles by dilution of lipids-octylglucoside mixed micelles. Chem Phys Lipids 66:93–109

Shew RL, Deamer DW (1985) A novel method for encapsulation of macromolecules in liposomes. Biochim Biophys Acta 816:1–8

Szoka FC, Papahadjopoulos D (1978) Procedure for preparation of liposomes with large internal aqueous space and high capture by reverse phase evaporation. Proc Natl Acad Sci USA 75:4195–4198

Szoka FC, Olson F, Health T, Vail W, Mayhew E, Papahadjopoulos D (1980) Preparation of unilamellar liposomes of intermediate size (0.1–0.2 μm) by a combination of reverse phase evaporation and extrusion through polycarbonate membranes. Biochim Biophys Acta 601:559–571

Vanlerberghe G, Handjani-Vila RM, Ribier A (1978) Les "niosomes", une nouvelle famille de vésicules á base d'amphiphiles non ioniques. Colloques nationaux du CNRS, Physico-chimie des composés amphiphiles, Bordeaux Lac, 27–30 Juin 1978, 304

Walter A, Lesieur S, Blumenthal R, Ollivon M (1993) Size characterisation of liposomes by HPLC. In: Gregoriadis G (ed) Liposome technology, 2nd edn, Vol. I. CRC Press, Boca Raton, pp 271–289

Welti R, Glaser M (1994) Lipid domains in model and biological membranes. Chem Phys Lipids 73:121–137

Index

vacuolar membranes, 32
vacuolar suspension, 32
Van der Waals forces., 38
vesicle transition, 232

yeast cells, 18, 42
yeast intracellular membranes, 25

Zeeman levels, 81
zonal centrifugation, 16
zymolyase, 103

Springer-Verlag and the Environment

We at Springer-Verlag firmly believe that an international science publisher has a special obligation to the environment, and our corporate policies consistently reflect this conviction.

We also expect our business partners – paper mills, printers, packaging manufacturers, etc. – to commit themselves to using environmentally friendly materials and production processes.

The paper in this book is made from low- or no-chlorine pulp and is acid free, in conformance with international standards for paper permanency.

MIX
Papier aus verantwortungsvollen Quellen
Paper from responsible sources
FSC® C105338

If you have any concerns about our products,
you can contact us on
ProductSafety@springernature.com

In case Publisher is established outside the EU,
the EU authorized representative is:
Springer Nature Customer Service Center GmbH
Europaplatz 3, 69115 Heidelberg, Germany

Printed by Libri Plureos GmbH
in Hamburg, Germany